THE
PERFECT
PROTEIN

THE

PERFECT
PROTEIN

The Fish Lover's Guide to Saving the
Oceans and Feeding the World

ANDY SHARPLESS, CEO OF OCEANA
AND SUZANNAH EVANS

FOREWORD BY PRESIDENT BILL CLINTON

RODALE.

For Beth

CONTENTS

FOREWORD

AMERICA IS BLESSED to have some of the world's most important fisheries within our national ocean zone. That's why, when I was president, I signed the Sustainable Fisheries Act into law, to help us to conserve as well as develop our fisheries. I embraced the goals of feeding people, maintaining the livelihoods of those who fish, and protecting the long-term abundance of the ocean. The law set up a system of management rules to make this happen by protecting our ocean habitat, limiting bycatch, and supporting scientific quotas. I consider it one of the top environmental achievements of my administration. It was one of the most successful environmental laws ever enacted, because it enabled the United States to become one of the first major fishing nations to make conservation a fundamental and effective part of the fisheries management process.

Enacting and implementing the law was a hard-fought, messy battle, one that continued in the courts for years after I signed it. It passed just a few years after the terrible collapse of cod in Canada and increasingly bad news about our own fisheries. My heart went out to the fishers and their families whose lives were wrenched by a rapidly declining catch. I was always impressed by those who supported conservation limits, despite the short-term impact on their livelihood, because they believed strongly that we needed to protect the future of fishing to save the way of life they love so much.

Their sacrifice paid off. Today, scientific reports show that America's fisheries, once headed toward decline, are now headed toward recovery. The results show that scientific ocean management works. In fact, the "caught in American waters" label is becoming an excellent indicator of the seafood one can eat sustainably. I hope this story, and our law, will be a source of encouragement, and even inspiration, for people all over the world. That's because most of the world's fisheries are still in terrible shape, and too many countries, including some that control most of the fish caught, do not yet have the policies necessary to manage their fisheries sustainably.

This issue is profoundly important to our future. We must have effective scientific management of the world's ocean fisheries, as this fine book points out, to provide the protein needed to feed a world population projected to reach 9 billion people by 2050.

The specter of ever-growing numbers of hungry people, especially malnourished children, hangs over our heads. Already, close to 1 billion people go to bed hungry. Through my Foundation, I work with farmers who cultivate small plots in developing countries to increase their production and their incomes. The results are impressive and, taken to scale, could do a lot to reduce hunger now and in the years ahead.

But we'll still need fish and their protein and other nutrients they provide to adequately deal with the challenge. I've never heard anyone else propose the simple solution Andy Sharpless and Oceana are making here: to replicate the success we've had in the United States by putting in place effective, conservation-focused, scientific fisheries management in the 25 countries that control most of the world's seafood catch. This is—relatively speaking—a practical, inexpensive, and quick way to make sure our planet has lots more nutritious food in the future, when we'll really need it.

I urge you to give this book, especially Andy Sharpless's proposals, a close read. I think you will enjoy the storytelling and find his ideas thought-provoking. I also commend Oceana for advocating for the oceans as a food source for people as well as for ecological reasons, and I hope more in the environmental community follow their lead. To allow everyone to thrive, our future requires that we find ways to keep our natural systems strong and abundant.

America's experience with ocean management proves that it does work. And I believe that, with proper enforcement, it will work in most countries. So I agree with Andy Sharpless: Let's save the oceans and feed the world.

—President Bill Clinton

CHAPTER 1

A Short Natural History of Seafood

Fish, amphibian, and reptile, warm-blooded bird and mammal—
each of us carries in our veins a salty stream in which the
elements sodium, potassium, and calcium are combined in
almost the same proportions as in sea water. This is our
inheritance from the day, untold millions of years ago, when a
remote ancestor, having progressed from the one-celled to
the many-celled stage, first developed a circulatory system in
which the fluid was merely the water of the sea.

—RACHEL CARSON, *THE SEA AROUND US*

IN THE SUMMER of 1935, the Inuit people of eastern Canada's remote Arctic
inlets were still living a subsistence lifestyle. Clad in sealskin leather, they
moved with the weather and the walruses, seals, and whales that they hunted.
The only plants they ate were the blueberries that grew during 2 months of the
year and sometimes the half-digested contents of caribou stomachs.

The fact that this nomadic hunter-trapper lifestyle persisted into the 20th
century in one of the world's harshest environments is incredible. Perhaps even
more unfathomable is the notion that the Inuits' restricted diet made them
among the healthiest people in the world. But in the early 1930s, the Inuits had

made contact with the modern world through the fur-trading Hudson's Bay Company, and the word was out that these tough Arctic people did not appear to suffer from diabetes, cancer, or arteriosclerosis.

Israel M. Rabinowitch, a chemist with McGill University, joined the Canadian government's annual supply trip to the Inuit communities in the summer of 1935 to find out if the stories about the hardy Inuits were true. Sailing aboard the HMS *Nascopie,* he traveled to four Arctic islands—Southampton, Devon, Ellesmere, and Baffin—and made dozens of visits to towns that were nothing more than temporary assemblages of a few sealskin tents.

In total, Rabinowitch examined 389 Inuits. While the population wasn't quite as disease free as reputation had it, they were remarkably healthy. He found no diabetes. Only one possible case of cancer. A few calcified arteries. Teeth worn down by chewing leather to make clothes and tents. "If there was a serious health problem amongst the Eskimos, [I] was not aware of it," Rabinowitch noted in a report published in the *Canadian Medical Association Journal* in 1936. He didn't even mention heart disease.

How could a people living in such harsh conditions not report a single heart attack?

During flush times, the Inuits ate 5 to 10 pounds of meat a day. Rabinowitch estimated that an average person ate 30 to 40 grams of carbohydrates, 250 to 300 grams of protein, 400 to 600 milligrams of cholesterol, and about 150 grams of fat *per day.* One hundred fifty grams of fat is the amount in 33 Twinkies.

Of course, not all fat is created equal. Inuit people weren't eating Twinkies. They were eating fresh, usually raw meat from marine mammals, fish, and occasionally caribou with their bare hands, sometimes ending meals covered in blood and blubber. At the time of Rabinowitch's study, science had not yet discovered why this high-fat, low-carbohydrate diet was good for their health.

Later studies would confirm the incredible vitality of the Arctic peoples. In 1980, Danish scientists compared rates of myocardial infarction, asthma, diabetes, multiple sclerosis, and other health problems among Greenland's Inuit with those of Danes. The data cleaved neatly, with the healthier Inuits showing little to no instances of the diseases compared with the "civilized" Danes.

The scientists didn't take into account one of the oldest uses of seafood—treatment of disease—because knowledge of the techniques had been lost to history for more than a century. In 1770, a woman suffering from severe rheumatism arrived at the Manchester Royal Infirmary in England. At the time, the standard treatment for rheumatism included "rubbing of her joints with cod-liver oil," as reported by Maurice Stansby in the seminal text *Fish Oils in Nutrition*. The poor woman received no relief, and after a year she inquired whether she could ingest the oil instead. She was brave: In the 1700s, the oil was obtained by pressing rotting cod livers, and it was an opaque, vile-tasting liquid. "Although the hospital had no confidence that this would have any effect, she was allowed to [ingest the oil]," Stansby wrote. Her symptoms disappeared.

Unfortunately for the woman, her doctors ascribed her good health to the changing seasons, and she wasn't given any more cod-liver oil for a year—during which her rheumatism flared more severely than before. The hospital relented and gave her more oil, only to see her symptoms disappear yet again.

After that, the Manchester infirmary regularly prescribed 1 to 3 tablespoons of cod-liver oil taken up to four times a day to treat rheumatism. A Dr. Thomas Percival wrote up the case for the *London Medical Journal* in 1783. But the difficulty of obtaining the oil, coupled with the truly revolting flavor (sometimes it was mixed with peppermint, although it's hard to imagine that helped), caused fish oil to fall out of favor, and it was forgotten.

Seafood's role in heart health was discovered only after those early-20th-century studies on Arctic peoples. Soon, other indicators emerged suggesting that seafood was helpful in avoiding heart disease. Norway experienced a steep decline in fatal heart attacks during the German occupation of 1941 to 1945. In these years, Norwegians could not obtain much in the way of meat, eggs, or whole milk, and instead began eating more fish, skim milk, and cereals. After the war, Norwegians returned to their red-meat diet, and the rate of heart attacks rose again. Similarly, scientists began to notice that the Japanese, who eat up to 13 times as much seafood as Americans, had much lower rates of heart disease as well. One study found that the Japanese were 20 times less likely than Germans to die of heart attacks.

One of the landmark studies on seafood consumption and heart health took place in the Netherlands from 1960 to 1980. Over those 2 decades, scientists tracked a group of adult men from the town of Zutphen who ate a consistent amount of fish throughout their lives. The result? The more fish the men ate, the less likely they were to die of heart disease.

After the results of the Zutphen study were published in 1985, the knowledge of seafood's role in heart health went mainstream. Now, just about every authority from the American Heart Association to the World Health Organization recommends eating seafood at least twice a week.

So what is it that makes seafood so healthy?

It has to do with its molecular structure. Seafood is the premium source for essential omega-3 long-chain fatty acids. The human body cannot generate these fatty acids by itself, so they must be consumed.

Since the 1980s, "omega-3" has been a nutrition buzzword, found everywhere from margarine labels to fad diet cookbooks. It's usually mentioned along with its relatives, the omega-6 fatty acids, which are derived from plant oils like soybean, corn, palm, rapeseed (canola), and sunflower. Omega-3 fatty acids are found in some plants, like walnuts, but the best sources are fish and seafood. They, too, ultimately derive their omega-3s from plants—the phytoplankton that support all ocean life.

• THE ABCs OF OMEGA-3s •

Omega-3 fatty acids are made of a string of carbon and hydrogen atoms. "Saturated" fatty acids hold all the hydrogen atoms their structures allow. Sometimes certain enzymes can remove pairs of hydrogen atoms. "Monounsaturated" fatty acids have one fewer pair of hydrogen atoms than the saturated versions. The loss of the hydrogen pair creates a double bond between the adjacent carbons. When more than one pair of hydrogen atoms disappears, fatty acids with two or more double bonds are created. These are known as "polyunsaturated" fatty acids. Omega-3 fatty acids fall into this category. The "3" in their name comes from the fact that their first double bond is three carbons away from the end of the chain known as the omega end. (The other terminus is the acid end.)

There are numerous kinds of omega-3s, but only three that we usually hear about in nutrition. Alpha-linolenic acid (ALA) is the only omega-3 found in plants, like walnuts, chia seeds, and flax. ALA isn't easy for the human body to use, but it's often ALA that's in foods like margarine, cereal, bread, and snacks whose packaging proclaims heart-healthy omega-3s are inside.

The more important nutritional benefits that we get from consuming omega-3s come from two other types of omega-3 fatty acids: eicosapentaenoic acid (EPA) and docosahexaenoic acid (DHA). These are found almost exclusively in marine sources and egg yolk, and yet they are critical to our health, having particular importance in fetal development and maintenance of brain, retina, heart, and immune system health.

• FISH IS (HEART AND) BRAIN FOOD •

The early study revealing that Greenland Inuits had little to no occurrence of diseases like myocardial infarction, asthma, and diabetes compared with Danes demonstrated two things: first, that *Homo sapiens* is not, on the whole, naturally given to these diseases, meaning that there are environmental forces at play. And second, that the high-fat diet of the Inuits correlated with their good health. The scientists found that the fish, seal, walrus, and whale diet of the Inuits—a diet that has nearly disappeared today because residents of remote communities now have access to as much processed, high-carbohydrate food as just about any big-city dweller—meant that about 14 percent of the fatty acids consumed by the Inuits were omega-3s. The Danes, meanwhile, were getting only about 3 percent of their fatty acids from omega-3s.

By the late 1980s, the link between omega-3s and heart health was firmly established. Scientists now agree that consuming omega-3-rich seafood two times a week can cut your chance of dying from a heart attack by 30 percent or more. Since then, scientists have explored other potential health benefits of omega-3s. Although not fully accepted by everyone, here are a few of the things that DHA and EPA omega-3 fatty acids may be able to do:

♦ Raise HDL, or "good," cholesterol

♦ Lower blood pressure

♦ Reduce the risk of heart arrhythmias

♦ Reduce the chance of forming blood clots that can cause heart attack or stroke

♦ Discourage the buildup of arteriosclerotic plaque by inhibiting the growth of smooth muscle cells

♦ Promote the resolution of inflammatory responses, resulting in faster healing from diseases and wounds

♦ Reduce an infant's chance of developing allergies

♦ Reduce the risk of depression, bipolar disorder, and suicide

♦ Protect neurons and improve brain cell repair, especially in conditions of brain injury or oxidative stress

♦ Lower the risk of Alzheimer's disease

♦ Ensure correct function and development of neurons, especially during fetal and infant development and in neuronal cell damage (e.g., stroke)

Our need for omega-3s begins before birth. A developing fetus absorbs fatty acids from its mother; DHA and small amounts of EPA are present in breast milk. The price for insufficient DHA and EPA is high. Without the brain-boosting omega-3s, babies have a greater chance of developing learning disorders. In the past, infants were often given a teaspoon of cod-liver oil, which is flush with vitamins A and D in addition to omega-3s, every day. But it wasn't until 2001 that manufacturers began fortifying store-bought formula with DHA.

All these data on the importance of omega-3s raise two important questions. First, how can we afford to fish out the world's oceans when they're such important sources of great nutrition? The United Nations says that, worldwide, 87 percent of all wild fisheries are already depleted or maxed out under intense industrial fishing pressure (happily, US fisheries are doing somewhat better). Imagine a world where children and pregnant women cannot access the omega-3s that are the building blocks for healthy brain and body development. As fish become scarcer, they will become more dear, too, leaving the neediest families in poor countries even more nutritionally disadvantaged, broadening the gulf between the haves and have-nots. How we avoid such a fate—and we can—is the rest of our story. But for now, let's focus on the second question raised by the potency of omega-3s: What quirk of evolution caused us land dwellers to develop such an affinity for a substance that comes from the oceans?

THE OCEANS PREDATE LIFE ON EARTH. Created by vast, thousand-year rains as Earth cooled, oceans—or really, one global, interconnected ocean—formed more than 4 billion years ago. The continents followed, born of Earth's erupting mantle. No one really knows where the first single-celled organisms appeared, but one common hypothesis places them in the hydrothermal vents of the deep sea.

While the continents remained barren, rock-strewn expanses, the oceans hosted blue-green algae that dominated life on Earth for more than a billion years. Through photosynthesis, the blue-green algae pumped oxygen into Earth's atmosphere until it reached a poisonous tipping point known as the Great Oxygenation Event, which killed off most life on the planet about 2.4 billion years ago. But the newly oxygenized atmosphere allowed other life forms to flourish. And so the oceans' dominant algae slowly gave way to multicellular organisms: sponges, jellyfish, squid, and trilobites. Fish, Earth's first vertebrates, appeared about 500 million years ago.

Algae likely colonized the shorelines long before any multicellular creatures climbed out of the waves. The first critter to venture up the beach 530 million years ago was probably a centipede-like creature scuttling along on dozens of feet. We can still see snapshots of evolution in the oceans: In 2010, scientists in Australia named nine new species of handfish with footlike protrusions instead of fins. Other ocean species are more like examples of arrested evolution. Some modern shark species have remained virtually unchanged over 100 million years, surviving even the extinction event that wiped out the dinosaurs.

For much of the last century, the accepted narrative for the evolution of early humanity has been what's called the *savanna theory:* The precursor of mankind, a species of great ape that lived in the trees of Africa, eventually moved down to the ground as the climate became hotter and drier. There, he scavenged for animals, berries, and fruits, eventually learning to hunt and walk upright. But life on the savanna, with its big cats and other predators, was more dangerous than it was in the trees. Marvin Harris, a prominent 20th-century anthropologist, spoke for many of his colleagues when he speculated that early humans were able to overcome the dangers of the savanna despite their lack of fangs or other natural weaponry by evolving the ability to use tools: "Those who brandish the biggest sticks are more to be feared than those who snarl with the biggest teeth." Coordination between hunters, and communication, speech, and culture, followed.

The savanna theory persisted with ample fossil evidence from places like the Awash Valley of East Africa, where the famous Lucy skeleton was discovered in 1974. Lucy was an example of *Australopithecus afarensis,* a precursor of *Homo sapiens* that showed bipedal posture and the small brain of an ape. But the savanna theory has been revised over the past couple of decades as geological evidence has shown that East Africa probably wasn't as dry as researchers once thought. It has been replaced in part by a scenario that suggests that we developed in a forested area, a much safer place than the savanna.

There's a third theory of human development that has a few ardent supporters. It aims to answer the question posed earlier: How did we evolve to become dependent upon marine-sourced food for the omega-3 fatty acids that are so critical for human health?

The first public proponent of a marine-based scenario of human evolution was Alister Hardy. When he spoke about his theory before a group of British scuba divers in 1960, he was a respected zoologist near the end of his career. Perhaps that's why he felt comfortable expressing his profoundly unorthodox opinion, one he'd mulled for 30 years in private, that the precursor of modern people had once had an aquatic phase of evolution.

Hardy had long wondered why humans, unique among primates, had a layer of fat beneath the skin like marine mammals, which were onetime land-based animals that returned to the oceans during their own evolution. He also marveled at *Homo sapiens'* superior swimming ability (for a land mammal at least) and at newborn infants instinctively knowing how to swim and dive. Perhaps most important in his theory was his idea that a watery phase may have allowed the early hominids to practice standing upright as they were buoyed by water, with the crowns of their heads protected from the sun by the only large patch of hair left on their slender, not-so-apelike bodies.

In a paper for *New Scientist,* a journal that never shied from publishing provocative theories that strayed off the beaten path, Hardy described his "aquatic ape" thusly:

I imagine him wading, at first perhaps still crouching, almost on all fours, groping about in the water, digging for shell fish, but becoming gradually more adept at swimming. Then, in time, I see him becoming more and more of an aquatic animal going farther out from the shore; I see him diving for shell fish, prising out worms, burrowing crabs and bivalves from the sands at the bottom of shallow seas, and breaking open sea-urchins, and then, with increasing skill, capturing fish with his hands.

Hardy placed his aquatic ape phase in a 2-million-year gap in the fossil record from about 5 million years ago. The reason for the gap, he theorized, was because hominids that died at sea would be much harder to preserve and uncover millions of years later.

But Hardy's aquatic ape was a highly controversial alternative to the prevailing savanna theory in the mid-20th century, and the lack of a fossil record hurt his credibility. Although Hardy's version was popularized in the 1970s by the writer Elaine Morgan, Hardy himself rarely wrote about it again and never provided biological or fossil evidence. The complex nature of the theory, which posited two previously unknown phases of human evolution (into the water and then the return to land), also kept it well beyond the fringes of acceptance. It's fair to say that any chance of mainstream acceptance of Hardy's theory died with him in 1985.

However, while contemporary scientists don't believe we were ever truly aquatic—and we agree with them—Hardy's controversial ideas did spark new thinking about an intermediate step in human evolution. Now, a few anthropologists are considering a new hypothesis that omega-3 fatty acids may have played a prominent role in the development of the modern advanced human brain. These omega-3s aren't available on the savanna or in the forest. But they are available in rivers, lakes, and, perhaps most important, the ocean.

Metabolic physiologist Stephen Cunnane is one of the few scientists starting to argue that neither the woodlands nor the savanna theory of evolution sufficiently explains how hominids could have evolved big brains and upright posture. Simply "needing" to become a tool user, or needing problem-solving or communication skills to hunt or avoid predators, can't spark big changes. The *biological opportunity* for this has to be there. It wasn't present in the savannas or woodlands because none of the other primates became human under the same environmental circumstances.

Cunnane describes a simple scenario: An australopithecine, perhaps a distant relation of Lucy, is scavenging for nuts and fruit. The drying climate means food has been harder to find, and eventually the hominid's wanderings bring him to the edge of a lake or river. Driven by hunger alone—because his tiny brain allows for little else—the hominid, for the first time, pulls apart a clamshell. He discovers it is edible—and delicious. Soon he brings other members of his clan to the water's edge, and they begin consuming clams, fish, and other shore-based foods.

The australopithecine doesn't know it, but he has stumbled upon a veritable all-you-can-eat buffet. The shorelines boast the world's greatest biomass, meaning its most plentiful food. Fish and crustaceans and mollusks like clams are easy to catch. Hominid clans could now settle by the water, a much safer place than the savannas, with less competition for food. This safer, less nomadic lifestyle was what allowed early hominids to develop tool-using skills. The copious nutrition from shellfish and fish also led to more rapid brain growth. Infants, too, were fatter and healthier—no other primates have fat babies. This high birth weight, relative to other primates, gave hominid babies what Cunnane calls energy insurance—protection for the healthy baby, and its developing brain, in the event of a short-term food shortage.

So, the theory goes, the early hominid clans that populated the shorelines were healthier and eventually smarter than their inland brethren. They protected their shores and pushed back competing bands. In addition to omega-3 fatty acids, seafood is rich in iodine, an essential trace element. For us, iodine deficiency is still a danger. It is the single greatest preventable cause of brain damage in the world, affecting mostly people who live in mountainous regions where iodine-rich seafood isn't available. The early hominids making a life on the shorelines were eating foods that were among the healthiest in the world for their evolving brains.

Alister Hardy didn't amass any physical evidence for his aquatic ape theory. But whatever the reality behind Cunnane's shore-based story of human evolution, human beings have an undeniable affinity for fish that dates back at least to the Middle Pleistocene epoch, about a 100,000 years ago. Archaeologists have unearthed 90,000-year-old harpoons near Lake Edward, on the border between Uganda and the Democratic Republic of the Congo. Scientists have also discovered evidence of shellfish consumption by early humans living on a coral reef near the shores of the Red Sea about 125,000 years ago. The Klasies River caves in South Africa are filled with fossil evidence that people ate oysters, limpets, and abalone more than 100,000 years ago.

Perhaps the greatest argument for the legitimacy of the shore-based scenario of human evolution is that we may still be in it. *Homo sapiens*—modern humans—is the only living species of the *Homo* genus. As the human population grew, it followed the paths of rivers and beaches. The five earliest written languages arose in cultures nestled by water. Rivers, lakes, seas, and oceans have dominated culture as sources of food, transportation, and mythology. We still thrive on a diet rich in the essential omega-3 fatty acids found in seafood, relying on it for basic health and development, and when we've strayed from it, we've paid for it with our health.

Seafood is also the only food with which we still have—mostly—the same hunter-trapper relationship as those early hominids cracking open clamshells. A shrimp or grouper from the Gulf of Mexico's warm waters can be flash-frozen and served for dinner anywhere in the world, and it's the only wild creature on the plate. Even the breaded fillets in McDonald's fish sandwiches come from the open ocean: Alaska's Bering Sea hosts enormous schools of pollock.

We may be evolutionarily disposed to enjoying seafood, but as our population has grown and grown, our collective appetite for wild-caught seafood has outstripped the oceans' ability to provide it. In order to keep up with demand, industrial fishing has wiped out much of the fish and shellfish that must have seemed endless to an australopithecine. The modern historical record is packed with examples: The Chesapeake Bay oyster, to name one, is now at less than a percent of the vast population found by John Smith's crew in 1607, the result of overfishing combined with pollution and disease. A valiant effort to restore the oyster fishery in recent years has been only moderately successful—and cost millions of state and federal dollars. Most species aren't so lucky; the industry wipes one out and moves on to another.

There's no question that we can't afford to decimate all wild seafood. Fish and shellfish are integral parts of our diets, and they should be. And they don't come with the massive baggage of industrial pork, poultry, and beef, animal proteins that produce tons of waste and pollution, destroy thousands of acres of land, use huge amounts of water, and are often too costly for the world's poorest people. The modern industrial agricultural system has mechanized food production in a way that's nothing short of awe inspiring for sheer effort. But we're paying a huge, often hidden price. And our planet may not be able to conceal the true costs of agriculture much longer.

CHAPTER 2

Reservation for 9 Billion, Please

It's obvious that the key problem facing humanity in the coming century is how to bring a better quality of life—for 8 billion or more people—without wrecking the environment entirely in the attempt. . . . The truth of the matter is that all the changes we make render the planet less suitable, not more suitable, for human beings.

—E. O. Wilson, Earth Day, 2000

IF THE DAWN of modern humanity was prompted by the availability of nourishing fat, then the 21st century may repeat that story, only this time through a fun-house mirror. An obesity epidemic looms today just as surely as hunger crises multiply. *Homo sapiens* is the dominant species on Earth. We've colonized nearly every corner of our planet, creating pockets of incredible wealth as well as valleys of famine and despair. A common way to look at the world today is through the lens of the haves and the have-nots. But let us suggest another way, harking back to the theory of the survival of the fattest: Humanity in the 21st century may be divided between two groups that we can call the "fats" and the "thins."

We have the technical ability to feed the world already, and quite fully. If you added up all the world's food and divided by the number of people on

Earth, each person would have 2,700 calories a day—plenty for survival. But, of course, famine still happens. Nearly a billion people on Earth are hungry, while another billion are overweight. Still another half billion are obese.

If we have the ability to feed everyone, why don't we? Obviously, economics and politics play enormous roles. Here's another, less well-known reason: More than half of the world's crop yields—mainly corn, rice, wheat, and soybeans—are used to feed livestock, not people. And most of the meat from the livestock is sold to people in wealthy nations. Jean Mayer, a prominent nutrition researcher and champion of school lunch programs, once estimated that we could save enough grain to feed 60 million people if we reduced meat production in the United States by just 10 percent.

That's a pretty good humanitarian argument for vegetarianism, and there's a similarly strong one advanced by environmentalists: Global agriculture uses 70 percent of the world's freshwater and is the single largest source of greenhouse gas emissions, primarily thanks to the resource-intensive production of livestock.

Despite the humanitarian and environmental benefits of vegetarianism, however, vast numbers of people will continue to eat meat. Indeed, as our population grows, and wealth grows with it, people seem to gravitate toward becoming more carnivorous. The world's population is growing faster today than at any time in history, passing 7 billion in 2011, just 12 years after hitting 6 billion. We're now on track to reach 9 billion by 2050. And the United Nations' Food and Agriculture Organization (FAO) says this will increase demand for food by 70 percent above today's levels. The demand for meat will skyrocket by 85 percent by 2030 and double by 2050 as more and more people pursue a Westernized diet.

A lot of this demand will come from the newly upwardly mobile. By 2030, the middle class is expected to number more than 3 billion worldwide, compared with 1.8 billion in 2010. Many of these middle-class citizens will be in China and India. China in particular has rapidly adopted a Western-style, meat-heavy diet. Half of the world's pigs live and die in China. In 1961, a Chinese person ate just 8 pounds of meat a year; by 2002, it was 115 pounds, and that number is due to grow by 40 percent by 2030. It's the same story from Albania to Vietnam, with some countries doubling or even tripling their per capita meat consumption since the 1960s. In the United States, consumption has held steady since that era, though we're number two at nearly 275 pounds of meat eaten per American. At number one, curiously, is Denmark, where citizens eat an impressive 321 pounds of meat per capita each year.

More meat in the diet means one thing: unhealthier people. Scientists have drawn a firm connection between meat consumption and disease. This is something we've known since the 1980 Inuit study, when the Danes, with their meat- and dairy-heavy diets, showed much higher rates of diabetes, heart disease, and cancer than the Inuits of Greenland did. And our declining health comes at a high price. In America, diabetes costs us $174 billion a year, while heart disease costs a staggering $666 billion, or $1 of every $6 spent on health care. One recent Harvard study of more than 120,000 Americans found that the more beef, pork, or lamb a person ate, the more likely he or she was to die during the study period. A daily increase of just 3 ounces of red meat corresponded with a 16 percent greater risk of death due to cardiovascular failure and a 10 percent greater risk of fatal cancer. The lead author of the study called the numbers "pretty staggering" in the *New York Times,* but he might have looked back to Rabinowitch's studies for a hint of how his results would turn out.

So we're not just creating a larger human population every year, we're also creating a growing population of meat-consuming and therefore unhealthier people who will cost the world untold billions for health care. And all the while, we are demanding more and more resources from the planet to provide the food they're eating. By 2030, 2.5 billion people could be overweight or obese thanks not just to meat, of course, but also to refined sugars and other processed foods.

The global food market has responded to this growing demand by producing meat faster, cheaper, and with less labor than ever before. It has done this through the rise of highly efficient feedlots where livestock such as cows and pigs are produced in huge numbers. This system of industrialized meat has created a lot of problems, however, both for the environment and for your health. Let's take a look at the costs of one of these, the animal behind the most popular meat on the planet: the humble pig.

WHAT IS THE PRICE of pork tenderloin? You might say $15 a pound for a really nice, fresh, special-occasion cut. How about 2 pounds of sliced bacon? Probably $10. Or your kid's baloney? A steal at $3.99 for a week's lunches.

The price you pay at the supermarket counter may make economic sense to the company that produced the pork you're taking home with you. But there are hidden costs with that pork product, too, costs that have grave environmental and social consequences. These costs, called externalities, are rarely included in the final sticker price. They are instead passed on to the municipalities that have to deal with the pollution and social problems associated with mass food production or are paid for with tax dollars in the form of government subsidies, which added up to $95 billion in the United States between 2001 and 2006 alone.

Paul Willis is the founding manager of the Niman Ranch Pork Program, a leading sustainable meat producer. The ranch contracts with family farms that raise livestock in harmony with the landscape, in stark contrast to the high-density feedlots that dominate industrial agriculture. As he put it, industrial feedlots are "bad for the animals, the environment, your neighbors, the water quality—honest to God, there's nothing good about it, besides volume."

But on that last measure, feedlots perform, and how. The United States is home to 60 million pigs. It was home to 60 million pigs in 1990, too, and those pigs produced about 15.4 billion pounds of meat. Now, the same number of pigs produce 21.7 billion pounds of meat in an industry that's worth $100 billion in the United States alone. This increase in production is largely due to the establishment of hog farms that raise the animals in enclosed warehouses. A single hog farm can house more than 10,000 pigs.

What are some of the problems associated with an industrial pig farm? Let's look to North Carolina for an example. Today, North Carolina hosts 10 million pigs—more pigs than people. But if you'd asked a resident in, say, 1985 if the Tar Heel State would become one of the global epicenters for producing pork, even processing 32,000 pigs a day at the world's largest slaughterhouse, you might have gotten a puzzled look in return. Until the late 1980s, the sloping coastal plain of eastern North Carolina was dotted with family farms. It took just half a decade for the state to become the country's second-largest pork producer, behind only Iowa.

The giant farms sprouted up with blinding speed for three reasons: The demand was there, the land was cheap, and the state was compliant, exempting the new companies from many environmental regulations. And since each hog produces four times as much waste as a person, those 10 million Tar Heel pigs create more sewage than the human residents of North Carolina, Pennsylvania, New Hampshire, and North Dakota *combined*. In the past, farmers had used manure as fertilizer on their fields, but today there are too few farms and far too many pigs for that synergistic system to work. Instead, much of this sewage is kept in open-air lagoons.

If the lagoons are compromised in any way—let's say, flooded by heavy rains—the results can be devastating. Until the BP oil spill in the Gulf of Mexico in 2010, one of the biggest pollution disasters in American history was a ruptured pig waste lagoon at Oceanview Farms in Richlands, North Carolina, in 1995. On a hot summer day, 25 million gallons of pig manure, urine, blood, and afterbirths flooded down US Route 258. The stinking river of sewage rose until it engulfed the wheels of cars. It eventually reached tributaries of the New River, killing thousands of fish in 17 miles of waterways.

But even when functioning correctly, open-air lagoons are constantly polluting the environment. They release vast amounts of greenhouse gases in a potent mix of 60 to 70 percent methane and 30 to 40 percent carbon dioxide, and they're one of the reasons why agriculture is the single largest contributor of greenhouse gas emissions. Livestock emit 37 percent of the world's anthropogenic methane and 65 percent of all nitrous oxide, two greenhouse gases that are vastly more efficient than carbon dioxide in trapping heat—23 and 296 times more efficient, respectively.

There are other problems with the pork industry that aren't unique to North Carolina. The eastern counties that contain most of the state's pig farms are among North Carolina's poorest, and that's not a coincidence: The farms lower real-estate values, attracting low-income, transient labor. Even the numbers of teen pregnancies and violent crimes showed spikes after industrial pig farms moved into one community.

And then there's the smell. Even on a calm, mild day, a visitor can catch a whiff of ripe air without being able to glimpse a farm, because they usually are set back from the road a quarter mile or more and screened by trees. But when it's really hot or when workers have sprayed the liquid effluent on fields as fertilizer, the smell can be overpowering to the point of inducing nausea. In a blistering article published in *Rolling Stone* in 2006, workers at a swine farm told the magazine that the smell goes away only when you grow new hair and skin, presumably while living somewhere far away from pigs.

And lastly, swine farm workers and residents of nearby towns can suffer terrible health effects. Thirty percent of workers at feedlots suffer from respiratory diseases like asthma. Living near a swine farm can affect your mental health, too, making you more depressed, angry, and tired. Perhaps most frightening of all, feedlots can be the sources of *E. coli* bacterial infections and powerful pandemics like the 2009 global swine flu outbreak.

These externalities have been the subject of reams of writing by respected thinkers like Michael Pollan, Eric Schlosser, and Wendell Berry. But meat has yet other costs that don't end at the feedlot. Each pig or cow or lamb has to be fed. It takes about 6 pounds of grain to produce 1 pound of pork. This is called the *feed conversion ratio*, or FCR, and it means you're eating acres of soybeans and corn, two of the main ingredients in pig chow, when you're eating meat. Pigs are in the middle of the pack when it comes to FCR efficiency. Poultry takes, on average, 2.3 pounds of grain to produce 1 pound of meat, and beef can take a staggering 13 pounds of grain.

You're consuming water, too, with that pork sandwich. It takes 3,500 gallons of water to produce a pound of pork. Just like with feed grains, pigs are somewhere in the middle for this particular resource usage: Poultry takes about 2,000 gallons per pound, and beef takes 2,500 gallons—the equivalent of running an average home faucet for nearly 2 days. This is going to be a major issue in the coming decades, with 64 percent of the world's population projected to be living in conditions of water scarcity by 2025. Freshwater makes up just 5 percent of the water on the planet, and agriculture already uses nearly a tenth of that. Currently, 1.7 billion people rely on aquifers that are being rapidly depleted, according to a recent study published in *Nature*.

Meat production is going to require more and more land and water to meet the greater demands of a growing population. After pork, poultry is the world's second-most-consumed meat. That industry is also going hog wild, so to speak: Between 1966 and 2006, US production of chicken meat rocketed upward by 500 percent. Beef is the third-most-popular meat, and American production of cattle is projected to grow from 31 million head of cattle today to more than 34 million in 2020.

It's a pretty simple equation: More meat = more water + more acres + more pollution.

And it means that the price you pay at the grocery store isn't the whole cost of the food. If agricultural companies actually had to include the costs of waste treatment, greenhouse gas pollution, deforestation, water use, taxpayer subsidies, and the other hidden costs in their prices, we'd see a huge difference at the market. A report by the International Sustainability Unit, a British think tank spearheaded by Prince Charles, revealed that $100 worth of Brazilian beef, for example, should actually cost in the neighborhood of $200.

These hidden costs mean that the world's agricultural resources are disappearing, a change that goes largely unnoticed by those of us in the developed world. But eventually, we're going to have to pay attention. Some scientists say we're already using 25 percent more resources than the planet can sustain. But as the land and water required to raise animals for meat become scarcer, it's the developing world that will feel the effects first. Globally, we're already experiencing the greatest volatility in food prices in 40 years, creating massive hunger shocks. In 2007, tens of thousands of citizens rioted in Mexico City as the price of tortillas skyrocketed by 400 percent. This was just the first in a series of 2007 and 2008 food riots that arose in Haiti, Mozambique, Morocco, Yemen, and other countries, all driven by vaulting prices. In the space of just 1 year, more than 100 million additional people became malnourished, tipping the world total to more than 1 billion hungry people.

Global warming, too, has played an insidious role. Major droughts around the world since 2006 have affected grain and cereal production everywhere from Australia to Argentina and Texas to Canada.

If we're just going to keep producing food that requires more land and more water all the time, how in the world are we going to feed a population that grows by 220,000 mouths every day?

IN THE MIDDLE of the 20th century, scientists met the challenge of a similar quandary posed by the booming post–World War II population. The 30 years of agrotechnological innovation that followed are often called the Green Revolution. Advances in fertilization, seed hybridization, and irrigation made the Green Revolution a history-altering success. The first country to implement it, Mexico, became a net exporter of wheat within a couple of years after introducing a hybrid seed that was hardier than its native counterpart. The effort went global. Between 1970 and 1990, the world's fields produced 2 percent more staple grains every year, keeping pace with the fastest human population growth in history.

But the Green Revolution has reached its limits. After 1990, yield growth rates fell by nearly half to just 1.1 percent a year. In 7 of the 8 years between 2000 and 2008, production of major crop staples actually fell *behind* consumption. This was due to a global drawdown of grain storage stocks, especially in China, after decades of stable food prices convinced governments that emergency stores weren't as vital as they once seemed.

Okay, so we need more land to grow more crops. But how much arable land is left on Earth? The answer to this seemingly basic question is a point of considerable debate. If you read some big-picture documents from the World Health Organization or the FAO, they count any place that has fertile soil and no big city built on top of it as "undeveloped arable land," including the rain forest in Congo. The political, ecological, and economic realities of clear-cutting the Congolese rain forest aren't included in the analysis. More than 80 percent of Earth's untilled arable land is in places like Congo. That's not really "available."

What we do know is that 30 percent of the planet's land is already used for grazing livestock, and most of that was once forested. Nearly three-quarters of the clear-cut jungle in the Amazon is now used as pasture for cows, chickens, pigs, and more. Think of that cost: Untold numbers of songbirds, insects, jaguars, and more vanished.

Agriculture and biodiversity have long been at odds. You could either save the forest and all its wild inhabitants or feed the people. This plays out time and time again in the modern world: cattle ranchers versus wolves in the American West, rice paddies versus mangrove swamps in Bangladesh, tea farms versus rain forests in Kenya. Conservation International has identified 25 of the world's top "biodiversity hotspots." Well, more than a billion people live in those hotspots, too, and more than half of them are poor and food insecure.

Should we have to decide between saving elephants and feeding malnourished people? This is an agonizing choice. Pitting these two global concerns against each other means that one will lose. For the most part, the loser has been biodiversity. Scientists say the current extinction rate is 50 to 500 times the average in the fossil record.

And recently, we've added another player to the mix: Both agriculture and biodiversity are now pitted against the biofuel ethanol, the production of which has contributed greatly to food price increases as crops are diverted from food to fuel. Two in 5 bushels of American corn are now used for fuel. And as land becomes dearer, both food and fuel prices will rise, again widening the gap between the fat and the thin.

But should we have to raze the world's remaining rain forests, mangroves, and other fertile environments to graze livestock and plant more wheat and corn and rice? Or dam and drain even more of the world's great rivers, some of which, even once-mighty ones like the Colorado and the Yangtze, are already so overutilized that they turn to trickles before arriving at the sea? The fact is that we're going to continue to exploit natural resources for food. The McKinsey Global Institute has estimated that in the next 2 decades we'll need to increase water and land availability by 140 and 250 percent, respectively, to meet the growing demand for food.

Here's what that could cost us by 2030: an additional 444 cubic miles of freshwater. That's the equivalent of the entire metro area of Baltimore submerged under more than ½ mile of water. We'll also slash up to 675,000 square miles of forest, an area the size of California, Texas, Montana, and Colorado combined. And lastly, we'll be pumping 66 gigatons of carbon dioxide into the atmosphere, a move that could help temperatures rise by 9°F over the next 80 years. Scientists have already said that even a rise of 3.6°F would be devastating for regions where poor smallholder farmers rely on rain-fed agriculture to feed their families. Considering that agriculture is the world's largest contributor of greenhouse gases, out-emitting even transportation, plus the fact that we're denuding forests that could help alleviate global warming, it seems that we're stuck in an unrelenting negative cycle.

But what if there was a healthy, animal-sourced protein that both the fats and the thins could enjoy without draining the life from the soil, without drying up our rivers, without polluting the air and the water, without causing our planet to warm even more, without plaguing communities with diabetes, heart disease, and cancer?

It's the one animal protein that's rarely mentioned in the endless reports about big agriculture and hunger crises. It's the protein that's healthiest for your body: low in cholesterol, brimming with brain-boosting omega-3 fatty acids and nutrients like riboflavin, iron, and calcium. It's one of the most ancient foods, and it's most likely the last wild creature that you'll eat, the last pure exchange between Earth and your dinner plate.

| The perfect protein

Imagine a world in which seafood, not pork, is the world's most eaten protein. You don't need a sprawling industrial landscape to feed the world with wild fish and shellfish. And you don't have to be rich to eat it. The world's poor— the thins—know this. A billion people on Earth already depend on seafood as their primary source of animal protein, and most of them are in developing countries. Four hundred million of the world's poorest citizens live in major fishing countries. Wild seafood accounts for 14 percent of the animal protein eaten around the world every day, and it does so without chopping down a single tree, without flooding fields and waterways with pollution, without emitting vast amounts of greenhouse gases.

Seafood is one of the world's truest renewable resources. It doesn't take millions of years to replace fish taken from the ocean, as it does coal from a mine. Fish are astonishingly fecund and resilient, so much so that as recently as the 1950s, people believed the sea to be inexhaustible.

As renewable as fish are, however, we haven't done the greatest job of stewarding the 71 percent of our planet that's covered by oceans. Only in the last couple of decades have we realized just how damaged the oceans have become. But one thing is clear: The oceans, and the multitudes of edible creatures they contain, have awed humanity with their bounty for centuries. It's this story that will show us how fish were once the key to new worlds and could be the doorway to healthier, less hungry lives for the 9 billion people on Earth in the 21st century.

CHAPTER 3

Shifting Baselines

The aboundance of sea-fish are almost beyond
beleeving, and sure I should scarce have beleeved it
except I had seene it with mine owne Eyes.

—Rev. Francis Higgeson, Massachusetts, 1630

In 1614, Captain John Smith landed on the little island of Monhegan,
a rocky scrap of land 10 miles from the coast of modern-day Maine. It was
6 years after his first landing in the New World, in a woodsy region named
Virginia after the Virgin Queen, Elizabeth. At that time, Smith had been a
26-year-old ne'er-do-well who had become so hated by his fellow colonials during
their voyage to America that his captain had planned to execute him upon landing. Six years later, however, Smith was leading his own cross-Atlantic expeditions
and had earned the right to name his discoveries. The name he chose for the vast
region that stretched between Newfoundland and Virginia was New England.

Smith and his men crossed the Atlantic in pursuit of gold, copper, and
whales. The minerals proved scarce, however, and the whales were difficult to
catch. So the captain set his sights on an easier quarry: fish. With just 15 men,
he hooked more than 60,000 cod in under a month.

And these weren't just any fish. The bottom-dwelling codfish was a
highly prized catch that had been fished in Europe for more than 1,000 years.

Cod is arguably the oldest cross-Atlantic market: It was in pursuit of the fish that the Basques of Spain first traversed the North Atlantic to get to untapped populations in unmapped regions. (The story, recounted in Mark Kurlansky's wonderful book *Cod,* has it that the sly Basques preferred to keep their rich fishing grounds a well-guarded secret, so the Vikings later got the credit for discovering Newfoundland.) By the time Smith landed in New England, Europeans had been sending fishing ships to Newfoundland for more than 100 years. They came back brimming with dried and salted cod, a bounty that one colonist called "better than the golden mines of the Spanish Indies."

The cod that Smith and his men hauled on board must have amazed them. They were monsters, three times as large as the fish in Newfoundland. The more Smith fished, the more he saw that the waters of New England brimmed with a ceaseless bounty.

> *You shall scarce find any Baye, Shallow Shore, or Cove of sand, where you may not take many Clampes, or Lobsters, or both at your pleasure, & in many places lode your boat if you please; Nor Iles where you finde not fruits, birds, crabs, & muskles, or all of them, for taking, at lowe water,"* Smith wrote. *"And is it not pretty sport, to pull up in two pence, six pence, and twelve pence, as fast as you can hale & veare a line? He is a very bad fisher, cannot fill in one day with his hooke & line, one, two, or three hundred Cods.*

Smith was not the only one to take notice of New England's marine bounty. The writings of colonists who made the journey in the 17th century are packed with descriptions of a vibrant sea so flush with life that it practically intruded onto the land. In 1632, an Englishman named Thomas Morton reported catching plaice by hand in ankle-deep water and schools of fish so densely packed in Massachusetts inlets that you could nearly use them to walk across the water at low tide. The halibuts were so plentiful, he wrote, that the fishermen ate just the heads and tails and threw the rest overboard. Clams that would have cost 12 pence apiece in England were fed to pigs in the New World. Lobsters were used as bait.

In 1661, one colonist was so moved by the ocean's bounty that he took to verse. Jacob Steendam's "The Praise of New Netherland" may not go down in literary history, but the quatrains aptly describe both the value of fish for sustenance and the sense that the supply was God-given and endless.

The lamprey eel, and sunfish, and the white
And yellow perch, which grace your covers dight;
And shad and striped bass, not scarce, but quite
Innumerable.

The bream, and sturgeon, drumfish, and gurnard;
The sea-bass, which a prince would not discard;
The cod and salmon,—cooked with due regard,
Most palatable.

The black- and roch-fish, herring, mackerel,
The haddock, mosbankers and roach, which fill
The nets to loathing; ane so many, all
Cannot be eaten.

And thus it happens here, that in the flood
Which, rolling from the Fountain of all Good,
O'erwhelms weak mortal man with royal food,
HE is forgotten.

You have to ask yourself why the colonists were so enthralled with the schools of menhaden so large that they flooded the harbor at Provincetown, the miles-long shoals of oysters lining America's coast, and the masses of alewife, an anadromous herring-like fish that swam upriver to breed in such masses that colonists could catch them by the thousand just by sinking a few rocks in their path. You'd think that, 400 years ago, the oceans all around the world would have been relatively pristine. After all, what could a global population of half a billion people lacking even steam-engine technology have done to the world's marine resources?

The answer: a lot. Colonists were astounded by the seemingly endless supplies of fish, crustaceans, dolphins, whales, and more because they had already forgotten what an abundant ocean looked like. Europe, with its 1,000 years of boats launching from fishing ports, had already begun to fish out its own waters. The codfish that John Smith hauled up were three times the size of the same fish in Newfoundland because New England's cod hadn't yet been subjected to a century of fishing pressure.

The Golden Age of Exploration and the centuries that followed it may as well have been called the Age of Exploitation. Even before the introduction of modern technology, we were methodically wiping out marine wildlife. Steller's sea cow, a huge, tubby, manatee-like marine mammal that swam in the North Pacific, was discovered by Europeans in 1741. It had been hunted to extinction by fur traders less than 30 years later. Fur traders had also pushed sea otters to the edge of oblivion by the early 1800s, and only legal intervention in the 20th century has kept them from going extinct.

The sea turtles of the Caribbean tell a similar story. The early explorers, including Christopher Columbus's son Ferdinand, wrote of waters so thick with turtles that the ships appeared to float on them. But today, the turtles' legacy, in many cases, persists only in the names explorers gave to the area's towns and islands, because the creatures themselves are nowhere to be found. Jeremy Jackson of the Scripps Institution of Oceanography in San Diego was one of the first ecologists to insist that we understand the oceans' former abundance. In 1997, he used historical accounts and catch figures to estimate the size of the green sea turtle population in the early 1700s, when turtle was the primary meat eaten in the British colony of Jamaica. He deduced that up to 39 million green sea turtles swam in the Caribbean Sea in the years before Columbus—a biomass 20 times larger than that of all the hoofed animals of the Serengeti. But the colonists made short work of the sea turtles in an assault on a par with the slaughter of the bison of the Great Plains. By 1800, the green sea turtle fishery of the Cayman Islands had collapsed. Only a few hundred thousand greens still exist in the Caribbean, just 1 percent of their original population. Today, all six species of sea turtles found in the Caribbean are in danger of extinction, along with the manatees that were also once abundant. As Jackson noted, even as a marine ecologist studying the region, "I have not even seen most of these large animals underwater for twenty years or more, despite thousands of hours scuba diving on and around coral reefs."

The sea cows and penguin-like great auks that went extinct or nearly so within a few generations of contact with Europeans were the sitting ducks, so to speak, of the marine world. These poor creatures had few defenses. Each one was slow and docile and offered either food, feathers, or fur to the captor. And when they were gone, you knew it. The North Pacific island beaches that once teemed with sea cows sunning themselves were bare. The Dry Tortugas were largely denuded of sea turtles. The last great auk mating pair was strangled by a trio of Icelandic sailors in 1844.

Fish, however, are a different story. Their disappearance is harder to see and easier to forget. Though colonists had already done a bang-up job of exterminating charismatic megafauna from the New World's waters, the first inkling of real trouble for America's fisheries didn't come until after the Civil War, with the sudden disappearance of a tough, voracious little fish called the bluefish.

Sleek, 2 to 3 feet long, with a forked tail and a scowling mouth gaping to reveal even rows of razor-sharp teeth, bluefish are rapacious predators that feed on large schools of menhaden, mackerel, and other baitfish. A bluefish feeding frenzy is an unforgettable spectacle. The predators arrive all at once. The sea foams as the bluefish rip apart their quarry, hardly pausing to swallow before tearing into another fish. They eat so fast and so indiscriminately that they create huge amounts of food for the seabirds that closely follow them to swoop and dive at the blood-streaked waters, and the crabs and other crustaceans who eat the chunks of fish scales and bones that drift to the seafloor. In one colonial account, fishermen caught 8,000 menhaden in 1 day in Massachusetts Bay. The next day, after the bluefish arrived, not a single menhaden could be found.

Few people eat bluefish today, as the flesh doesn't keep well. But the fish's bloodlust makes it a prime recreational catch because it takes bait easily and then puts up a good fight. In the early decades of America's settlement, however, people were less choosy about their seafood. Bluefish were the primary fish eaten from New Jersey to Massachusetts in the summer, when the big schools could be counted on to arrive offshore. Fewer than 15 men could load a boat with 1,500 bluefish in a day. A fixed net could catch 100,000 of the fish overnight.

That is, until the summer of 1871. After a subtle decline over a decade, suddenly only half of the bluefish that had appeared in New England the year before showed up. The fishermen were baffled. Bluefish had been a reliable and lucrative catch for decades. Now, boats with twice as many men on board came back with less than half the fish they had just a year earlier.

Bluefish wasn't the only catch that appeared to be dwindling, although the still waters left in the wake of their disappearance presented the clearest evidence of their absence. New England fishermen complained that they were catching fewer sea bass, striped bass, scup (also known as porgies), and blackfish.

At the time, there were two main kinds of fishing in America: hook-and-line fishing from boats, and passive fixed-gear fishing, which used nets and traps left to soak in the water to snatch anything that passed through. The fixed gear tended to be owned by businessmen and landowners, while the boat-bound fishers were on the other end of the economic scale. It added up to resentment along class lines as the line fishermen became convinced that the fixed-gear technique was responsible for their empty hooks.

In 1871, spurred by the reports of dwindling fish catches, the assistant secretary of the Smithsonian Institution volunteered to undertake the United States' first scientific fisheries assessment. Spencer Baird was a zoologist who had already made his mark with contributions to ornithology and biology. He was the rare scientist who was also a skilled lobbyist. With three boats, $5,000 obtained from Congress, his Smithsonian salary, and a few volunteers, Baird began the US Commission of Fish and Fisheries in 1871 at Woods Hole, Massachusetts.

The report that Baird filed the next year concluded that fish catches were indeed dropping in New England. In some cases, the situation was dire. In

Nantucket, fish catches had dropped by 75 percent. Of Buzzard's Bay, one of Massachusetts's oldest fishing grounds, he wrote, "The whole business of fishing was pretty nearly at an end."

In order to save the fisheries, Baird recommended that the fixed gear be removed during spawning seasons. He stopped short of saying they should be banned outright. It didn't matter. Neither Massachusetts nor Rhode Island, the sites of his studies, adopted his recommendations.

Baird's fisheries commission ultimately settled on a different tack to stave off the depletion of America's seafood. Instead of focusing on overfishing, it turned its attention toward hatcheries. Baird pioneered the shipping of salmon roe by rail from coast to coast in an attempt to prop up the Atlantic salmon supply, but the scheme was a failure. (Today, wild Atlantic salmon are commercially extinct. Any "Atlantic" salmon you see on a menu has been farmed.) And Baird fell victim to a logical fallacy that is still used today by some in the fishing industry: He blamed the bloodthirsty bluefish for the falling fish catches, despite the facts that bluefish had played the same role in the marine ecosystem since time immemorial—and that the disappearance of the bluefish was one of the mysteries he had set out to solve. "Indeed, I am quite inclined to assign to the blue-fish the very first position among the injurious influences that have affected the supply of fishes on the coast," he wrote.

But in more ways than one, Baird was well ahead of his time. He unflinchingly noted that the commercial fishermen were working toward a single, injurious goal: "to obtain the largest supply [of fish] in the shortest possible time, and this has involved more or less of waste, and, in some cases, reckless destruction of the fish." He also warned of a coming scourge that would change the face of fisheries forever: the trawl, a type of fishing gear that was already in common use in Europe.

"Should this engine of destruction come into general use on our coast and add its agency to those already referred to in connection with the pounds and weirs [traps and fixed nets], the diminution of the supply may continue to go on in a vastly greater ratio than ever," Baird wrote.

He didn't know how right he was.

TRAWLING REVOLUTIONIZED FISHING. For millennia, humans had been catching fish by net, trap, spear, and hook. The first bottom trawls were 20-foot nets weighted by stones and lead in the closed "cod" end. The front end of the net was held open by a wooden or steel beam. Pulled by a sailboat going with the wind and tide, the trawl raked the seafloor and scared flatfishes like flounder, halibut, and sole into the net.

The first mention of a bottom trawl in historic literature, dug up by Callum Roberts for his definitive history of fishing impacts on the oceans, *The Unnatural History of the Sea,* was prescient: It was a complaint. In 1376, English fishers wrote to King Edward III to request his intervention in the use of the "wondyrechaun," a weighted net dragged along the seafloor to snatch up anything in its path.

> *And that the great and long iron of the wondryechaun runs so heavily and hardly over the ground when fishing that it destroys the flowers of the land below water there, and also the spat of oysters, mussels and other fish upon which the great fish are accustomed to be fed and nourished. By which instrument in many places, the fishermen take such quantity of small fish that they do not know what to do with them; and that they feed and fat their pigs with them, to the great damage of the commons of the realm and the destruction of the fisheries, and they pray for a remedy.*

The bottom trawl was so unpopular among hook-and-line fishers in Europe that they succeeded in staving off the new and damaging technology for centuries. Its use was banned in several countries and even made a capital offense in France in the 16th century. But as Roberts theorizes in his book, there may have been another reason why the trawl didn't catch on: It caught such volumes of fish—sometimes making the net so heavy it could not be hauled up onto the boat—that the fishers couldn't sell the catch before it spoiled. It wasn't until the advent of railroads and the widespread exportation of ice from Northern Europe that the enormous bounties realized by trawling could actually be utilized. By the 1860s, just 30 years after the world's first steam passenger service started, more than 100,000 pounds of fish were transported by rail in England each year. With a new market of seafood consumers beyond the coasts now reachable, the number of British trawlers increased sixfold in 2 short decades, to more than 800 in the early 1860s.

By the numbers alone, the era when fishermen resisted the advent of the trawl was over. But many still complained that, by ripping up the seafloor and crushing the oyster beds and rocky expanses that were the homes and sources of sustenance for fish, the trawls were killing the goose that laid the golden egg. They clamored so loudly that a royal commission was set up by the British government in 1863 to investigate the complaints. But the commissioners rejected the fishermen's concerns outright, instead claiming—contrary to the facts—that the trawls actually *fostered* life by furrowing the seabed like a plow turning dirt in a field of wheat.

One of the members of the royal commission was Thomas Henry Huxley. A biologist sporting voluminous sideburns, Huxley had earned prominence as one of the earliest and most vocal supporters of Darwin's views on evolution. But during two royal commission investigations into trawling (the second one undertaken in 1883), Huxley embraced the role of skeptic. He sneered at the fishermen's complaints about dropping catches and dismissed their on-the-water accounts.

By the time the second commission on trawling was launched in 1883, fishing had been revolutionized again by the addition of steam engines. The steam engine allowed trawlers to reach the ocean floor regardless of the conditions, further intensifying the destructive power of the gear. The classic beam trawl was also increasingly joined by the otter trawl, a modification that keeps the front of the net open with steel or wooden doors that gouge out huge furrows and send up an opaque cloud of mud, rocks, seagrass, and anything else that might be in the path of the net. The benefit of the otter trawl for fishers was that it could catch groundfish like cod, not just the flatfish scared up by the beam trawl. In some cases, the otter trawl pumped up fish catches by 50 percent literally overnight.

Fisherman after fisherman testified before Huxley and the commissions, pleading for regulation of the trawls. But Huxley was unmoved. In the same year that the second commission was conducted, he gave the inaugural address to the International Fisheries Exhibition in London. To Huxley, the riches of the British and European seas were still as plentiful as those of the virgin coast of New England had once seemed to John Smith.

> *I believe that it may be affirmed with confidence that, in relation to our present modes of fishing, a number of the most important sea fisheries, such as the cod fishery, the herring fishery, and the mackerel fishery, are inexhaustible. And I base this conviction on two grounds, first that the multitude of these fishes is so inconceivably great that the number we catch is relatively insignificant; and, secondly, that the magnitude of the destructive agencies at work upon them is so prodigious, that the destruction effected by the fisherman cannot sensibly increase the death-rate.*

Five hundred years of complaints against the onslaught of the trawl meant nothing to our doubting Thomas. So convinced was he that the oceans' bounty could not be reduced by the hand of man that he called any regulation of the seas "useless."

A naturalist named Walter Garstang set out to discover over the following decade if Huxley's claim—that "in relation to our present modes of fishing" the sea was inexhaustible—was true; to do so, he launched the first study of the trawl fisheries of England. His findings should have been sobering. Despite the introduction of steam and otter trawls purported to make fishing more efficient than ever, every single trawl fishery on England's east coast saw an overall decline in catches from 1889 to 1898. In Lowestoft, England's east-ernmost fishing town, overlooking the North Sea, cod catches in 1898 were just *2 percent* of what they had been in 1883.

Today, a fishery is often defined as collapsed when catches drop below 10 percent of their historic maximum. Thanks to the brutal efficiency of trawls, England's fisheries were well on their way to oblivion. The willful blindness of Huxley and his contemporaries set an ominous tone for the 20th century. It was in this century that we would fully realize the extent of our collective fishing might.

SPENCER BAIRD HAD BLAMED the messy, voracious bluefish for decimating fish populations in New England, but he had the foresight to warn against the trawler as an even more destructive force. Like the bluefish, trawlers are incredibly wasteful predators, destroying everything in their paths. But unlike the bluefish, which create a buffet for the scavengers traveling in their wake, thereby helping to maintain the existing food web, bottom trawlers leave only rubble, making it difficult for the marine environment to recover to its natural state. Bottom trawlers destroy 4 to 16 pounds of marine life for every fish they catch. This waste, known as bycatch, can include other species of fish (including sharks), sea turtles, dolphins, octopuses, corals, and more—anything that has the misfortune to be caught in the path of a rumbling trawl.

Today's trawlers use much the same technology as in Baird's time. But the difference is twofold. First, in scale: The world's largest supertrawler, the *Atlantic Dawn,* debuted in 2000. It isn't just a fishing ship. It's a huge floating factory. Weighing 14,000 tons unloaded and stretching longer than 1½ football fields, the *Atlantic Dawn* can store 18 million servings of frozen fish in its hold. The ship's otter trawl is supersized, at 200 feet wide and 40 feet tall: A 747 jet could fly through the metal doors. The weighted net is so heavy and powerful that it can sink to the bottom and shove aside 25-ton boulders as it chases down fish.

The second difference is the digital technology. It's been a long time since anyone could catch flatfish by hand in shallow waters, as colonists described with such heady enthusiasm. Modern industrial fishing ships are equipped with satellite technology, seabed-mapping software, sonar, radar, GPS devices, and more tools that transform the ships into highly sophisticated fish-seeking missiles.

Most trawlers aren't nearly as massive as the *Atlantic Dawn*. Just 1 percent of the world's fishing fleet can be called supertrawlers. But these few massive ships catch a huge portion of the world's seafood.

Meanwhile, in coastal regions and in seas like the Mediterranean, you're more likely to see boats hardly bigger than a large speedboat hauling the otter trawl's telltale steel doors. But what these ships lack in tonnage, they make up for in number. Together with the supertrawlers, they rake a seafloor area twice the size of the continental United States every year. Spencer Baird would be loath to visit New England's remaining fishing ports today. The trawler fleet there fishes an area the size of Rhode Island, Massachusetts, Connecticut, New Hampshire, and Maine *combined*.

Trawlers are the flagships, so to speak, of a global fleet that has expanded exponentially since the end of World War II. The naturalist Walter Garstang reported that fishing effort had more than tripled on England's shores in the decade he studied the nascent trawling industry, and yet catches were dropping. And now we have added to the weaponry other specialist fishing gear like the purse seine, which targets fish at the top and middle of the ocean's water column. The seiners encircle schooling fish with 1,000-yard nets before dragging them into the hold. Without laws to protect marine mammals, it is purse seiners that are the most likely to drown dolphins and porpoises chasing schooling tuna.

This story has repeated itself again and again on a global scale since the early 20th century. Improved technology has allowed fishing fleets to search farther, deeper, and longer for fish. This expansion is not driven by some unspoken desire to conquer the oceans, like summiting Mount Everest or hiking Death Valley. It's because we've already laid waste to the marine wildlife that was easiest to catch. We started with the slowest and most trusting seabirds and marine mammals (sorry, Steller's sea cows!) and are now in pursuit of the most elusive fish in the world's remotest underwater places.

Take the Patagonian toothfish, for example. Twenty years ago, no one thought to eat this slow-growing deepwater fish, which was found only in the recesses of underwater canyons in the high seas of New Zealand, Antarctica, and southern Chile. But with its tasty and easily cooked meat—and a name change to Chilean sea bass—the once-obscure toothfish suddenly found itself on the menu at upscale restaurants all over the United States in the 1990s. Just as quickly, deepwater longliners started a gold rush on the previously unexploited fish. Only a few years later, toothfish numbers crashed, and it was put on the Monterey Bay Aquarium's red list for overexploited species.

When the modern industrial fleet turns its attention to vogue fish like the Chilean sea bass, the resulting carnage can be breathtaking. Sharks are another devastating example. In the last few decades, the market for shark fin soup has expanded exponentially with the growth of the upwardly mobile Chinese middle class. Once a delicacy reserved for the elite, shark fin soup is now proudly and frequently served at weddings and other important events as a sign of wealth. The shark fin cartilage is tasteless and nutrition free, so any flavor in the clear soup comes from the broth; it is only the shark's symbolic weight that keeps it on Chinese menus.

The fishing industry has responded by waging an all-out war on sharks. Since the fins are so valuable, fishermen haul sharks on board, slice off the fins, and throw the sharks overboard to bleed to death, like poachers kill rhinoceroses just to take their horns. Up to 70 million sharks die this ignominious death every year, carnage equally inhumane and ecologically unsustainable. And because it's extremely difficult to identify a shark species by the fins alone, fishermen get away with slaughtering the most vulnerable sharks, from the fearsome great white shark to the harmless filter-feeding basking shark.

Big predator fish like sharks have suffered huge losses from fishing since the 1950s, when the commercial fleet became fully mechanized after World War II. But just as with the depleted populations of sea turtles brought to light by Jeremy Jackson only in the late 1990s, we've been late to notice sharks' dwindling numbers. As late as 1954, two top academics published a book called *The Inexhaustible Sea* that echoed Huxley's perennial optimism that the oceans were simply too large for humanity to ever empty of life. And yet the late Ransom Myers and his colleague Boris Worm showed in a landmark study in 2003 that many big predator fish—like tuna, swordfish, marlin, cod, flounder, and halibut—had been reduced to less than 10 percent of their 1950 numbers. And as we've seen, by 1950, many of those fish had already been suffering from intense fishing pressure for some time.

One reason we didn't notice the world's failing fisheries was because it was disguised for years in the official data. If you looked at United Nations figures for global fish catch in the 1990s, it showed a steady increase every year for the entirety of its data set. But Daniel Pauly and Reg Watson, fisheries scientists at the University of British Columbia, were skeptical. China, one of the world's behemoths when it comes to seafood consumption, was consistently reporting jumps in catch every year since the 1980s—a claim that flew in the face of on-the-water reports of diminishing catches. So Pauly and Watson redid the data in 2001, correcting China's overinflated claims and accounting for vast natural fluctuations in the world's biggest fisheries in South America. With the data corrected, the statistics suddenly showed a different story: Global fish catch was not rising. It wasn't even holding steady. Instead, it had peaked in the late 1980s at about 90 million tons and has been in jagged decline ever since, even though we were searching for the most obscure fish in the deepest recesses of the oceans with megaships like the *Atlantic Dawn*.

There's another reason we haven't noticed the creeping desertification of the once-vibrant seas. We simply fail to remember. Pauly coined the term *shifting baseline syndrome* in 1995 to describe our collective amnesia when it comes to what constitutes healthy oceans. It's a syndrome with a long pedigree. John Smith, accustomed to the smaller fish common in England's North Sea as a result of centuries of overfishing, showed it with his amazement at the massive codfish abundant in New England's healthy waters.

The concept of shifting baselines was adeptly demonstrated by a doctoral student's project in 2009. Loren McClenachan from the University of California's Scripps Institution of Oceanography pored over hundreds of photos of trophy fish caught in Key West over the last half century. The photographs showed the biggest catches of the day, like groupers, sharks, and sawfish, hung on a wooden dockside rack. In the 1950s photos, huge fish competed for space on the rack, draped one over another. By the 1980s, the fish were small enough to be displayed in nicely contained rows. And by the 2000s, the trophy fish of the day were barely bigger than what you might find in a well-stocked koi pond. Yet today's recreational fishers tacking their best catches to the board and posing for pictures feel just as proud as their predecessors.

It's also hard to talk about shifting baselines when the average American, European, or Japanese person can walk into any grocery store and see huge volumes of seafood from around the world filleted and ready for sale. The supply in just the frozen-foods section of your average Safeway seems inexhaustible! The truth is that most of this fish is imported from far-flung places. It's been a long time since New Englanders could reliably eat huge amounts of cod, herring, and oysters from their own shores. But our appetite for seafood only grows. The supertrawlers, with their onboard fish-processing centers, have made it possible for us to get flash-frozen fish from virtually anywhere in the world, often at the cost of the world's poorest, who stand by as their governments sell the right to fish in their national waters to the highest bidders.

So what's a progressive-minded seafood lover to do? Is it possible to choose healthy, sustainable seafood?

CHAPTER 4

The Consumer's Dilemma

Buying the right kind [of seafood] seems to require an
advanced degree in endangered species. Fish shopping,
in short, is not for sissies, and it's fraught for anyone
with an environmental conscience.

—MARK BITTMAN, "THE DISH ON FISH"

IT'S ONE OF THE MARVELS of modern American life that you can walk into
virtually any grocery store and buy seafood from around the globe any day of
the week. You can stand still and cast your eye across Arctic cod, Chilean sea
bass, Hawaiian mahi-mahi, and Vietnamese shrimp in one glance. And it's a
testament to how normal this has become that it's the *local* catch that's often
presented as something notable. Today's special: seafood that would have been
your only option before the advent of flash freezing and globalization.

This familiar display of marine abundance is one of the fundamental
drivers of ocean depletion because it disguises overfishing, allowing us to continue to purchase disappearing fish in blissful ignorance. If American, European, and Japanese people encounter daily confirmation of the ocean's bounty
in their grocery stores, how can they ever be expected to believe that the oceans
are running out of fish? The wealthy nations are the ones that can pay the most

for these "limited-edition" fish, so the last fish in the world will be served on a plate in New York, Tokyo, London, or Dubai.

And because these rich nations are vastly more influential than those in the developing world that provide so much of the fish we eat, the mirage of fish abundance on store shelves hurts our ability to correct (and see and understand) overfishing. This is not merely a theoretical or abstract point. Already, most American seafood is not American at all. The United States imports more than 90 percent of its seafood. There are a couple of reasons for this, beyond our willingness to pay among the world's highest prices. First is our sheer appetite. In 2011, America surpassed Japan to rank second in the world in total seafood consumption, behind just China. We ate, collectively, 4.7 billion pounds of fish and shellfish, or about 15 pounds per person.

Surprisingly, this is about the same total tonnage that we catch. So why do we import so much fish? Seafood is an international commodity, and Americans have a certain sensibility when it comes to what we like to eat. We prefer large, top-of-the-food-chain predators with flaky, mild-flavored flesh: tuna, cod, grouper. Some of these fish can be found in US waters. Some of them used to be here but got wiped out by trawling and other forms of industrialized fishing by the late 20th century. And some of them were never here. Instead, the fishing industry sought them out, like the Chilean sea bass that's hauled up from the coldest underwater canyons to be sold on dinner plates a hemisphere away.

The industry has to keep plumbing those unknown depths to find the next lucrative catch because it's already aggressively fished most of the world's fish populations. Nearly 9 in 10 of the world's 600 fisheries are already fully exploited, overexploited, or recovering from depletion, according to the United Nations, which is generally pretty conservative in its estimates. A "fishery," by the way, is defined by a number of factors that include species, geography, and the type of gear used. The Mid-Atlantic scallop trawl fishery, for example, may share waters with a smaller diver scallop fishery, where individual divers gather the same species of scallop by hand. But in the eyes of regulators, these are two different fisheries, leading to a complex quilt of rules in many of the world's ocean waters.

Democracy has a lot of virtues, but its response to a slow-growing and mostly silent problem is not one of them. Lawmakers are generally pushed to action only when faced with a highly visible calamity. "You never want a serious crisis to go to waste," President-Elect Barack Obama's chief of staff, Rahm Emanuel, famously said in 2008 when discussing the new administration's plans for dealing with the financial collapse. But the "useful crisis" that the oceans need is not going to come to the powerful and wealthy democracies of the world. Instead, fish will be flown in from farther and farther away, and the seafood shelves will stay full. Prices will rise but slowly and without drama. People with money will spend it. Fish will quietly become a luxury item.

Crisis-driven policy making means that many Americans assume that the biggest problem facing the seas is pollution, because that's what makes the news. It's true that oil spills are a grave threat, as we were reminded in 2010 in the Gulf of Mexico. Chemical runoff from big agricultural operations has generated algae blooms that create anoxic marine dead zones around the world. The buildup of toxins like mercury in the flesh of fish is a real health issue, prompting formal warnings from the US Food and Drug Administration (FDA) that children and women of child-bearing age should limit their fish consumption. Even the so-called Great Pacific Garbage Patch, a massive slurry of degraded plastics in the middle of the ocean, has received the full crisis treatment from the media. Oil spills and garbage patches included, however, no single industry has altered the oceans like industrialized commercial fishing.

If the democratic process seems unlikely to address a problem that's silent and nearly invisible, at least to the world's powerful countries, you might wonder if educated and impassioned consumers could step into the gap. The path for the individual citizen sounds simple enough: Buy only sustainable seafood. But anyone who's been to a grocery store lately knows it's more complicated than that.

FIRST, WE HAVE TO DEFINE our terms. What is *sustainable*? Like its cousin *natural*, it can be slippery. There is no official government definition, and any fish supplier can slap it on labels. For our purposes, sustainable seafood is a wild fish or shellfish that is harvested at a scientifically determined rate that allows the population to rebuild itself each year. This means that the next year's catch can be just as large or larger than last year's and that this repeated annual catch is as high as possible without risking next year's spawn.

The people who manage fisheries generally agree on this rule in principle, but there is nevertheless some debate over the precise definition of optimal fishing levels. The conventional definition is called *maximum sustainable yield*, or MSY. Conservationists point out that this should be treated as a limit, but commercially minded managers view it as more of a target. This means that, in practice, the amount of seafood caught often exceeds the MSY even when managers start the season with it in mind.

For typical shoppers, however, something like MSY isn't the definition of sustainability that comes to mind. There's a second, slightly better-known component to a well-managed fishery: Sustainable seafood should be harvested without any detrimental effect on the marine environment and other wildlife. You might not know the term *bycatch*, but you're aware of its effects if you've ever bought a can of dolphin-safe tuna. As we talked about in the previous chapter, bycatch is the fish and wildlife that's caught alongside the targeted seafood species. Trawling is responsible for up to 50 percent of the world's bycatch, but other industrial methods, from longlines studded with thousands of hooks to football-field-size purse seine nets, can also catch and kill everything from seabirds to octopuses to marine mammals.

But back to that dolphin-safe label. Fishing ships can locate schools of yellowfin tuna by following dolphins, which eat the fish just as people do. As a result, when a fleet encircles a school of tuna with purse seine nets, the fishers also catch the dolphins. Videos of hundreds of dolphins thrashing and drowning in nets surfaced in the 1980s, leading to consumer outrage and a boycott. Realizing they had a major public relations problem, the world's three largest tuna companies agreed to stop buying tuna caught this way. Official bans on tuna caught with high levels of dolphin bycatch followed in a number of countries.

Bycatch remains a major problem in many other fisheries, however. Gulf of Mexico shrimp trawlers, for example, catch and drown sea turtles as they drag the seafloor in search of shrimp nestled in the mud. All six species of sea turtles found in American waters are protected under the Endangered Species Act, but the shrimping industry has won concessions to allow it to kill them. In the early 2000s, conservation groups including Oceana succeeded in getting the government to require improved turtle excluder devices on shrimp trawls. These are escape hatches that allow sea turtles to swim out of the nets, reducing mortality by about 90 percent. Still, shrimpers kill hundreds of endangered sea turtles every year. That's not something most people think about when they're grilling up Gulf shrimp.

It wasn't until the late 1990s, a decade after the dolphin-safe tuna campaign, that the first large-scale consumer boycott related to overfishing took place. The Give Swordfish a Break campaign, organized by SeaWeb and financed by the Pew Charitable Trusts, was probably the first time most Americans had considered the reality of overfishing. Swordfish, once hunted by harpoon, were almost exclusively caught with longlines by the 1970s. These lines can be dozens of miles long and sport thousands of hooks that are sometimes lit with glowing lures to attract swordfish. The doomed boat *Andrea Gail*, featured in *The Perfect Storm*, was a swordfish longliner, hauling up a quarter million dollars' worth of the fish before the wind kicked up.

By the time the *Andrea Gail* succumbed to the waves in 1991, North Atlantic swordfish were in trouble. The average size of the fish had plummeted from 266 pounds in the 1960s to less than 100 pounds. Big and slow growing, female swordfish don't reproduce until they hit about 150 pounds, so this was a serious problem. The federal government continued to set quotas too high, against the advice of scientists, and the commercial industry denied that the fish were on the decline. So the Give Swordfish a Break campaign hit them right in the wallet with a nationally coordinated consumer boycott.

The result was both a marketing and a conservation success. Hundreds of chefs dropped swordfish from their menus, followed by major hotel chains and cruise line companies. President Bill Clinton voiced his support for a ban on catching swordfish smaller than 33 pounds. Wild Oats, a regional grocery chain that has since been absorbed into Whole Foods, stopped stocking swordfish altogether.

Within 3 years, the federal government had closed fishing in more than 100,000 square miles of key breeding areas for the fish. Now, North Atlantic swordfish populations have recovered beyond the target level.

Give Swordfish a Break introduced the concept of overfishing to many Americans. And it had a very real effect. It produced a healthy swordfish population by giving fish consumers the tools they needed *as citizens* to goad policy makers into action. Its success lay not merely in convincing people not to eat the fish but also in helping to produce new policies.

But more than a decade later, national consumer campaigns directed at one species of fish are rare. Instead, we've gotten a patchwork of measures seeking to help shoppers buy sustainable seafood. Probably the best known is that of the Marine Stewardship Council (MSC), which certifies specific fisheries according to its own definition of sustainability. Certified fisheries feature the blue MSC sticker on their products, which consumers can use to identify which fish to purchase. Currently, about 7 percent of the world's wild seafood is sold with an MSC sticker on it, adding up to about 170 fisheries, with another 300 in the process of obtaining MSC certification. Unfortunately, while the approval process is open and quite detailed, some conservation organizations have grown suspicious. The MSC tends to certify any fishery that has submitted an application. And the MSC's certified fisheries include bottom trawlers and longliners, which most marine scientists regard as inherently incompatible with sensible and sustainable ocean management.

A different guide comes from the Monterey Bay Aquarium's Seafood Watch. This color-coded system ranks and labels fisheries as green, yellow, or red based on in-depth assessments by the aquarium's scientists. Since it was started in the late 1990s, Seafood Watch has come in a folded pocket guide that describes dozens of species of fish. Since 2009, it also comes in a downloadable application for both the iPhone and Android. Now you can stand at that seafood counter and consult your cell phone whenever you have a question about sustainability. If you're lucky, the fish will be listed. As good as it is, though, the Seafood Watch app isn't yet all-inclusive.

Still, we have to acknowledge that only a small fraction of shoppers care or know to ask if their grouper fillet comes from a healthy population or whether it was pole-, net-, trawl-, or line-caught, and whether any other animals died in the pursuit of that fish. People are busy. As a practical matter, asking them to take the time to check on all their fish choices is probably asking a bit too much. This is especially true when the information on the card conflicts with the evidence of apparent abundance in the store. Indeed, the Give Swordfish a Break campaign may have worked in part because it was not systematic and comprehensive. Its single-species focus and the simplicity of its call to action—boycott the fish—allowed busy people to easily remember and act on it. This signal was also clear enough to reach the policy makers. But the message policy makers should glean from citizens' use of guides like Seafood Watch is a complex and politically difficult one.

The practical person will conclude that we need to eliminate the pressure on the consumer to make the sustainable choice. You should be able to walk into a grocery store or restaurant and order seafood armed only with the knowledge that anything you buy is good for the oceans. This should reasonably be the purview of the government, which after all is responsible for ensuring the safety and reliability of the food supply. Yet no government anywhere in the world has taken responsibility for ensuring that only sustainably caught fish are sold to its citizens. So the first steps that might eventually help catalyze this action by the government will probably be taken by the private sector. Fortunately, in the last few years, some retailers have been making strides in that direction.

A NATIONAL LEADER in grocery-store seafood sustainability is Whole Foods Market, the upscale grocer with a progressive reputation. In 1999,

Whole Foods teamed up with the Marine Stewardship Council to offer certified fish in its stores; in 2010, the company also partnered with the Monterey
Bay Aquarium and Blue Ocean Institute to rate fish in its stores that weren't
certified under MSC. At the same time, the company made a bold pro-
nouncement: It would stop selling all red-rated seafood by Earth Day 2013.
It beat that deadline by a year, phasing out a dozen remaining species like
gray sole, sturgeon, skate, imported wild shrimp, and trawl-caught Atlantic
cod in 2012.

These days, most Whole Foods locations get a shipment of seafood six or
seven times a week. On a typical Tuesday in the company's store in Silver
Spring, Maryland, just past the Washington, DC, line, the seafood delivery
arrives on a pallet at 6:00 p.m. It holds a few packing crates brimming with ice
and whole, gutted trout and salmon alongside bags of littleneck clams and scallops. Collectively, the pallet's fish traveled thousands of miles: the tilapia from
Ecuador, the catfish from North Carolina, and so on, and all ended up together
at Whole Foods' Mid-Atlantic seafood warehouse in Landover, Maryland,
before being distributed to the regional stores. The next day's special is
Chesapeake Bay scallops, a Maryland classic. Dozens of them arrive in a clear
plastic bag. They smell sweetly of a distant ocean.

Most of this seafood will be sold within the next day or two. But for now,
it's spot-checked by a Whole Foods employee, who inspects the fish for bright
red gills and shiny, clear eyes—the signs of a fresh catch. Once the order is
cleared, the seafood is placed in a cooler behind the counter until it's time to
restock the display.

The seafood counter itself is a sight to behold. Fillets in shades of pink,
white, and orange line the display as blackboards with crisp white lettering
explain the species and the sustainability rating of each fish. "Catch the reel big
news," one says in the company's signature ingratiating tone. "NO more wild-
caught seafood from red-rated fisheries!"

The Silver Spring store was one of the company's first to go sustainable.
When executives asked for a volunteer region to serve as a pilot for Whole

Foods' transition away from red-rated seafood, the first hand up belonged to Kevin McDade. He's the seafood buyer for Whole Foods' Mid-Atlantic region, which encompasses 41 stores from New Jersey to Virginia. An enthusiastic guy with a crew cut and a megalodon tooth hanging around his neck, McDade acknowledges that Whole Foods' initial transition toward more sustainable seafood wasn't widely embraced by customers.

"People said, 'Why did you take that fish out of there?'" McDade said. "I told them, 'I got another fish you can buy. It's going to be very similar to what you had, and let me tell you why this is what we're buying.' Our customers, at the beginning they didn't understand it."

Fishermen were even more reluctant to embrace the new regime, which meant expensive gear changes at best and a loss of income at worst. Several years ago, sensing the direction the company was headed in, McDade began working with New England cod trawlers to encourage them to change how they fished so they could keep selling to Whole Foods. While the hope was that there would be a switch to pole-catching fish (a much less damaging way to fish), the fleet ultimately switched to gill nets, which have their own set of issues, including large amounts of incidentally caught fish and wildlife that can end up ensnared in nets. Fishermen complained to the *New York Times* in April 2012 that they were being hurt by a corporate move that amounted to little more than green-washing in their eyes. McDade acknowledged that the shift was a tough one. "We got hammered," he said.

"We're buying everything we can buy from them," McDade said. "So other people are going to look and say, 'Hey, these guys got solid customers. They're buying this stuff like crazy.' That's what we want, so other people will change, and then the dominoes start falling."

The East Coast cod fishery, which had been rated red by the Blue Ocean Institute, moved to yellow. If the cod fishery continues using gill nets, it's unclear how the fishery will ultimately reach McDade's goal: He says he'd like to see yellow-rated seafood disappear so that only sustainable, green-rated fish and shellfish are on his shelves. "What we're trying to do is have a sustainable-seafood program and not rape the oceans of all the goodies and leave nothing but a blank, empty chest," McDade said.

WHOLE FOODS MARKET has started on a path toward grocery store sustainability, but it represents a small, and admittedly elite, portion of the market. It has just more than 300 stores, compared with Kroger's 3,800. Encouragingly, other chains have made steps toward sustainable seafood, too. Wal-Mart announced in 2011 that it would transition to only MSC-certified seafood, and by 2012, nearly three-quarters of its seafood met that requirement. (As mentioned, though, the value of MSC certification is questionable.) BJ's Wholesale Club has said it will sell only sustainable seafood by 2014. Greenpeace named Safeway its top-rated sustainable-seafood grocery chain in 2012 in its annual *Carting Away the Oceans* report. Over the 6 years the report has been issued, the 20 retailers whose seafood practices were analyzed stopped selling 67 species of fish rated red by Greenpeace. (Confusingly, there are multiple red lists; Greenpeace's is independent of Seafood Watch's and Blue Ocean Institute's.) One of the discontinued fish was orange roughy, a long-lived deep-sea fish, highly vulnerable to fishing pressure, that disappeared from 13 of the chains that once sold it. Now just 5 American grocery chains carry the heavily overexploited fish, which has the unfortunate double whammy of also coming from a trawl fishery.

A number of notable chefs have also joined the sustainable-seafood movement in recent years. Hugh Fearnley-Whittingstall, a renowned British chef, teamed up with UK grocery chain Sainsbury's in 2011 to get Britons to try something that wasn't one of the heavily exploited species. Fully 80 percent of the seafood consumed in the United Kingdom comes from what are called the "big five" species: cod, haddock, tuna, salmon, and prawns. With the exception of prawns, these are all top-of-the-food-chain fish that require a lot of resources and time to grow to maturity, which is rarely a characteristic of a resilient and sustainable species.

In nearly 400 stores across the United Kingdom, on June 17, 2011, Sainsbury's offered customers the opportunity to try less exploited, lower-on-the-food-chain coley, pouting, megrim, rainbow trout, and mackerel—for free. The effort seemed to take hold in the British consciousness: Sales of the big five species dipped slightly during the campaign, while the stores sold an extra 50 tons of the promoted sustainable species.

"If enough customers ask for something like coley or pollock in store, then soon enough those sustainable fish start to appear," said chef Jamie Oliver, who joined Fearnley-Whittingstall in the campaign. "You can't beat the call of consumer demand."

Fearnley-Whittingstall even hosted a series on the UK's Channel 4 dedicated to revealing how the oceans are being ruined for Europe's seafood. In *Hugh's Fish Fight,* Fearnley-Whittingstall adopted the role of a seafood version of muckraking filmmaker Michael Moore, detailing his efforts to get an interview with a reluctant executive at Tesco (the United Kingdom's largest fish seller) to present a video of an admission from a Tesco-supplying purse seine fisherman that the nets regularly caught dolphins and sharks.

Perhaps more remarkably, Fearnley-Whittingstall tackled the decidedly unsexy issue of "discards" in not just an episode of the television program but also a policy-focused international campaign. Discarding fish is the practice of dumping fish overboard because they aren't the boat's targeted species or because European Union regulations won't allow them ashore for one of a number of confusing, fishery code–related reasons. Discarded fish can include prized species like cod or haddock or lesser-known but still edible fish. They're all thrown away like trash. No one really knows how many fish are discarded in EU waters or around the world, but it's a common practice. An estimated half of the fish caught in the North Sea are thrown away. Smartly, Fearnley-Whittingstall allied himself with the fishermen rather than blaming them for laying waste to the oceans. This collegial approach allowed him to film fishermen throwing perfectly good fish over the gunwales with regret in their eyes— a memorable image for the millions of Britons watching on Channel 4 and paying a pretty penny for ever-scarcer codfish and chips.

Six months after the first series aired in 2011, sales of the sustainable fish promoted during *Hugh's Fish Fight* had risen by up to 50 percent at Tesco. The retailer reported steady sales of cod, however, so people weren't exactly abandoning the more exploited species. But it's still an important step toward getting consumers to consider eating species whose removal has fewer negative effects on the oceans.

Fearnley-Whittingstall took the Fish Fight campaign to policy makers with a push to enact a discards ban in the European Union. By 2012, he had gathered nearly a million signatures on an online petition and won a promise from the European Union to phase out discards over the next 5 years—an incredible triumph of consumer activism. Just as the Give Swordfish a Break campaign showed, the true measure of success for a consumer-based sustainable-fishing campaign is policy action. Fearnley-Whittingstall showed that it can be done, even with an issue as wonky and unknown as discards.

In the United States, numerous other chefs have taken up the mantle of sustainability. Barton Seaver, a chef in Washington, DC, launched the city's first sustainable-only seafood restaurant with the opening of Hook in 2007. The next year, Seaver opened Tackle Box, a casual lobster-shack establishment that serves fried tilapia and bluefish along with clams, mussels, and oysters. Some chefs joined the movement, at a cost to their businesses. In Seattle in 2009, Hajime Sato became one of the world's only Japanese sushi chefs to eliminate unsustainable seafood from his menu. He lost a significant chunk of his business in the first 6 months. When he visited friends in Japan, he got an especially negative response. "People told me I was destroying the culture because I could not eat this or that," he said. "I said that's bullshit. If we don't do this, we're not going to have any fish left."

On Long Island, Sam Talbot found that most customers didn't really think about the provenance of the fish on their plates at the Surf Lodge, a hotel restaurant he helped open in 2008. He didn't face the negativity of Sato's clientele—more of a mild indifference that he hoped he replaced with awareness, one plate at a time.

"Once people realize they're doing something responsible that they weren't even aware of when they walk in the door of the restaurant, they start to realize that this guy is really trying to do the right thing, and they look up other chefs who are doing that," Talbot said. "I think that response is wonderful."

For many chefs, sustainable seafood meshes with the common desire to serve fresh, local ingredients. Talbot's dedication to sustainable seafood is part of a larger philosophy on food. "I'm constantly looking for the most ecoresponsible source," he said. "And usually when you find those people, their products, their fish, their vegetables, their honey, it's far superior to everything that's out there."

There's no doubt that a growing number of chefs have embraced sustainable seafood. It seems like a natural outgrowth of the slow-food and locavore movements that have been percolating for a decade or more. It's an ethos epitomized by, well, Whole Foods, which sells to consumers who are really, really, *really* interested in the story of their food.

And yet Whole Foods customers and chefs who study the provenance of the fish they serve every night represent just a tiny percentage of seafood eaters in America. The truth is that most of us don't think too hard about the Filet-O-Fish sandwich at McDonald's, the spicy tuna roll at our local sushi joint, or the T.G.I. Friday's grouper special. Somebody should, though. With shocking regularity, the fish you eat isn't the fish you think it is. No one is safe, even those shopping at upscale grocers like Whole Foods, which struggles with this issue just as much as midlevel grocery chains. It's a whole new kind of crime, and it's happening right under your nose: seafood fraud.

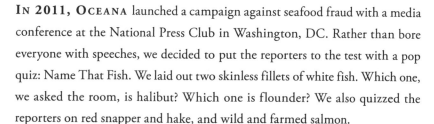

IN 2011, OCEANA launched a campaign against seafood fraud with a media conference at the National Press Club in Washington, DC. Rather than bore everyone with speeches, we decided to put the reporters to the test with a pop quiz: Name That Fish. We laid out two skinless fillets of white fish. Which one, we asked the room, is halibut? Which one is flounder? We also quizzed the reporters on red snapper and hake, and wild and farmed salmon.

You probably won't be surprised to learn that most of the reporters flunked the test. Most consumers don't spend enough time with fish to really know what they're eating. That's what makes seafood fraud so easy to commit. Unfortunately, neither the government nor most retailers are doing a very good job of following your fish from bait to plate, and that's the other reason why seafood fraud has become rampant.

When a Copper River salmon destined for Whole Foods' seafood case arrives at the dock in Alaska, the chain's local buyer enters it into an electronic database. The purpose of this system is to track where fish come from. Copper River would like to use that data to trace where its fish is ultimately sold. Some of the local companies that Whole Foods buys from are going even further. The J.M. Clayton Company, which sells jumbo lump crabmeat from the Chesapeake Bay, marks each tub with the number of the individual picker who plucked off the crab's flesh and packed it. If a customer complains about too many bits of shell in his purchase, the news gets back to the picker—and when you're talking about Whole Foods, where even parking can have the intensity of a blood sport, it isn't hard to believe that this does happen. Whole Foods itself needs to go further. The company has yet to require traceability on all of its seafood. Safeway, meanwhile, has stepped up to the plate; it has announced company-wide measures to follow seafood from hook to fork starting in 2015.

Currently, most seafood isn't nearly that traceable, especially when it's crossing international boundaries. The United States imports more than 90 percent of the seafood we eat, and that can involve a maze of distribution and packaging centers quite literally around the world. Wild salmon that was caught in Alaska, for example, can go to China for packing and be a "product of China" by the time it arrives in your grocery store.

And here's the unsettling part. Only a tiny fraction of that seafood is actually inspected by anyone. The FDA is supposed to inspect foreign seafood, but it actually looks at only about 2 percent of the millions of metric tons of seafood that arrive in the United States every year from more than 130 countries, according to a Government Accountability Office report. And of that 2 percent, just a sliver—0.05 percent—is specifically investigated for fraud.

This matters for a couple of important reasons. The first is that seafood fraud—labeling one fish as something else—contributes to overfishing by both obfuscating the true state of a fishery and contributing to the belief that over-fishing isn't happening—how could it be when the desirable red snapper, for example, is on menus everywhere? Or you could think you're buying a sustainable fish when you're actually getting something off the red list.

And when processors or distributors swap out the fish they're selling, well, they're not usually upgrading you. The desirable species are generally replaced with lower-quality fish. (The sole exception may be a worldwide chain of elite sushi restaurants that surreptitiously sold Atlantic bluefin tuna in its London outlets without labeling it as such, despite the fish's status as a massively over-fished species in many parts of the world. Eating Atlantic bluefin tuna is like eating a rhinoceros to most conservationists. It took DNA testing by Greenpeace to reveal that Nobu was selling Atlantic bluefin in 2008.)

Beyond profit, however, fraud can have major health implications, and not just for people with allergies. In 2007, 600 people fell ill in Hong Kong after eating what they thought was Atlantic cod. What they were really eating was escolar, an oily fish with such a nasty reputation for its gastrointestinal effects that it's been nicknamed the Ex-Lax fish. This is a fish you definitely want to avoid, although you may not be able to. In a 2012 investigation, the *Boston Globe* found escolar was being sold as white tuna, super white tuna, and albacore in multiple sushi restaurants and Asian grocery stores throughout the city, despite the fact that you couldn't buy escolar back in Japan if you wanted to—the fish is banned there. The paper also found that Boston-area sushi restaurants were regularly serving tilapia, a cheap (although sustainable) farm-raised fish, and calling it red snapper. "Not because we are trying to trick," one restaurant manager told the *Globe*. "We're doing it how everybody does it."

In the course of a 5-month investigation, the *Globe* conducted DNA tests on 183 fish samples from 134 restaurants, grocery stores, and markets. Nearly half—48 percent—were sold under a species name different from what they actually were, and almost all of those came from fish markets and restaurants, not grocery stores.

The *Globe*'s results track with the findings of an Oceana investigation in Los Angeles in 2012, in which 55 percent of the 119 seafood samples our scientists collected were mislabeled. Nine of 10 sushi samples collected weren't what they were claimed to be, and every single fish with the word *snapper* in the label was mislabeled according to federal guidelines. And another study conducted by Oceana in 2012 in New York City found that 39 percent of the 142 samples collected were mislabeled, including 13 different types of fish that were sold as "red snapper."

Fraud, unlike sustainability, is one of those issues where the consumer has essentially no power. You can ask a higher-end restaurant to show you a whole fish before you order a fillet, but you'll have to be expert enough to identify a species by sight. Because fraud is much easier to commit when the seafood has been processed in some way, you can buy raw, unprepared fish from a seafood counter rather than frozen fish that's been marinated or breaded. But the bottom line is that change here needs to take place at the government level. The FDA currently inspects imported seafood at a much lower rate than the USDA inspects beef and poultry, despite the fact that we eat less seafood as a whole and yet get sick from it more often. (Seafood is responsible for 15 percent of documented foodborne-illness outbreaks.) The ease of DNA testing these days will eventually make fraud harder to commit, but in the meantime, we're working to get the government to do a better job of protecting the health of both consumers and the oceans. The federal government needs to require seafood traceability that makes the process from bait to plate transparent and verifiable, provide consumers with more information about the seafood they purchase, and keep illegally caught fish out of the US market. That would give the purchaser the ability to know when seafood was caught, where it was caught, how it was caught, and what its species is. This is not as difficult as it sounds. The Europeans are phasing in a version of this system right now. The United States, hopefully, will be next.

THE FOOD WRITER Michael Pollan famously boiled down his philosophy into seven words: "Eat food. Not too much. Mostly plants." When people ask us which seafood is sustainable, it's hard to give such a pithy response. Red lists, or green lists for that matter, aren't static. Healthy fish populations vary geographically, and fisheries can be defined by gear, location, and time of the year. Bycatch is a hidden cost that most people don't think about.

But if you really pressed us for it, this is what we might say: "Eat wild seafood. Not too much of the big fish. Mostly local." The United States has some of the best-managed fisheries in the world because we've put good laws on the books to stop overfishing, protect nurseries, and reduce bycatch. As a result, many of our fisheries are beginning to rebound. And domestic seafood doesn't rack up a large carbon footprint like fish flown from points across the globe. Large predator fish like Britain's big five are already heavily exploited and slow to recover; meanwhile, lower-on-the-food-chain, quick-reproducing fish like the mackerel and coley promoted by Hugh Fearnley-Whittingstall are much more resilient to fishing pressure. And wild fish are in concert with nature, while farmed fish combat it. Just how ugly fish farming can get is our next topic.

Can We Farm Our Way to Abundance?

Farming as we do it is hunting, and in the sea we act like barbarians.

—Jacques Cousteau

HALF THE FISH and shellfish eaten around the world today don't come from the open oceans or meandering rivers. They come from pens, tanks, and racks and were bred in captivity or captured and stuffed full of feed to fatten them up for market. This is aquaculture, and it's the world's fastest-growing food sector.

Aquaculture is nothing new. The Chinese started raising freshwater carp more than 3,000 years ago, and fossil records and ancient literature describe the fishponds of India and Assyria as well as the penchant of wealthy Romans for cultivated oysters. Indonesians raised milkfish and mullet in shallow ponds called *tambaks* about 700 years ago in the first documented examples of marine fish farming. But for most of mankind's history, freshwater species have dominated attempts to farm fish.

It will come as no surprise that the belief in the endless bounty of the oceans led most entrepreneurs to ignore the possibility of raising saltwater fish. And yet at the same London International Fisheries Exhibition where Thomas Huxley famously called the oceans' bounty "inexhaustible" in 1883, a few other scientists were laying the groundwork for aquaculture. One of them took millions of salmon eggs back to New Zealand; another introduced trout farming in South Africa. Scientists at the Woods Hole Oceanographic Institution in Massachusetts had already successfully bred and released cod in 1878. But most of these efforts went to supplementing wild populations, not to creating an independent economy of raising and selling farmed marine fish. Sport fishers were the main beneficiaries of hatchery programs in the early 20th century. Truth be told, the need for large-scale commercial fish farming just didn't exist. Even in 1950, the world's human population was still less than 3 billion, and fish were plentiful and cheap.

All that changed when the big commercial fisheries started to decline in the second half of that century. Suddenly, fish farming started to look like a growth industry. And unlike livestock and grains, both of which include just a few big-dollar species, aquaculture may be the most diverse farmed food sector on the planet. It encompasses every pocket of the food web, from kelp and seagrasses grown for biofuel and cosmetics, to scallops and oysters, to Atlantic salmon and the most valuable fish in the sea, the rare and mighty bluefin tuna. By the numbers, aquaculture is a roaring success. Per capita, we now have access to 17 pounds of farmed seafood, up from just 1.5 pounds in 1970. The industry has grown by 8.8 percent a year, outpacing that of the world population, and is worth nearly $100 billion. Today, two-thirds of the salmon and 9 out of 10 shrimp eaten in the United States are farm raised.

Only about a third of the world's farm-raised seafood, however, is marine. That's because the most commonly farmed fish in the world is carp, the favored species of the world's leader in fish farming, China. Sixty-four percent of the world's farmed fish is grown in China, and most of that is carp raised in inland ponds.

There is clearly an opportunity for expanded marine aquaculture to help shore up our dwindling ocean resources. But it has to be done right. All too often, it's been troubled by the same problems found in the world's land-based industrial feedlots. Let's look at two examples of aquaculture, one good and one not so good, to demonstrate the opportunities and costs of farming the oceans.

SALMON IS ONE of the world's most popular seafood species. For conservationists, it's also one of the most frustrating. Wild salmon have been virtually exterminated from East Coast rivers in the United States, while salmon runs in the western United States have become increasingly unpredictable. In 2008, federal regulators canceled the entire California salmon fishing season for the first time in 160 years after 90 percent of the chinook salmon in the Sacramento River failed to return to spawn. Fishing wasn't allowed to resume until 2010, and even then the low numbers of salmon returning to the river's delta perplexed scientists. No one can pin the salmon's dire numbers on a single cause. Rather, it's probably a longtime mix of overfishing, habitat destruction, pollution, damming, and climate change. The collapse has cost California somewhere in the neighborhood of $1.4 billion in lost economic opportunity and 23,000 jobs. In Alaska, wild salmon populations are in much better shape. Regardless, wild salmon is a scarce and therefore premium product. You can spend up to $30 a pound for its heart-healthy, omega-3-packed flesh.

It's no wonder that people would try to replace the wild version with a farmed equivalent.

Salmon is an anadromous fish, which means that it's born in freshwater and travels to the ocean before returning to rivers to spawn. This migration is one of the great spectacles of the natural world, with salmon thrashing and wrestling their way upriver, back to the places of their hatching. Scientists aren't sure exactly how salmon know which river to return to after several years in the open ocean. Some theorize that the fish have an acute memory for smells; others suggest that they have electromagnetic sensors or use the position of the sun as a guide. Whatever it is, it makes the salmon a powerful, muscular fish; it has to be in order to survive.

A farmed salmon, however, doesn't need any of these mysterious homing functions or a natural athleticism. Instead of a female salmon sweeping out a depression in the riverbed for her eggs, which are then fertilized by a passing male, farmed salmon eggs are spawned in freshwater tanks at onshore hatcheries. (This blurs the line between farmed and wild fish; releasing land-bred salmon fry has been one of the primary functions of the US Fisheries Commission since the days of Spencer Baird. Today, up to 80 percent of the United States' wild salmon get their start in hatcheries. For our purposes, hatchery-raised released salmon still count as wild.) And instead of battling their way upstream, morphing the actual physiology of their muscles as they leap against the tide, farmed salmon swim sedately in a pen, jostling their neighbors and bingeing on a constant stream of pellet feed. Because wild salmon garner their pink flesh from the krill they eat, the pellets contain shades of pink and red dye the farms can select from a color palette to avoid ending up with salmon that have unappetizing, gray flesh. The differences are more than skin deep, however. Foodies often complain that farmed salmon lack the firm muscle tone and strong flavor of their wild brethren, which those fish earn not just from their diet but also through their Herculean life cycle of coursing from river to ocean to river again. And farmed salmon can have half as many omega-3 fatty acids in their flesh, despite containing up to twice as many grams of fat.

Still, farmed salmon are a relatively healthy source of protein. Given the volume of salmon lovers versus the number of wild salmon in the world, on its face, farming salmon doesn't seem like the worst idea. But the industry has been plagued with problems. Salmon are cold-water fish, so the first salmon farms were established in Norway and eventually expanded to other Northern European nations and Canada. But one place where farmed salmon have made the biggest splash, so to speak, is 7,000 miles from their native northern rivers, in Chile.

Salmon aren't native to the Southern Hemisphere, but they couldn't have found a more suitable adoptive home than Chile's pristine, pollution-free fjords. The inlets of the country's southern regions are sheltered from the elements and maintain a perfectly chilly temperature of about 50°F. Fed by plenty of rainfall and the hundreds of rivers that sidestep down the Andean foothills, these waters are close to a mirror image of those in Alaska and western Canada.

For decades, Chileans had tried to introduce salmon the old-fashioned way, with hatcheries and wild releases. Even the Peace Corps helped breed and release thousands of coho and chinook salmon in Chilean rivers before their work was interrupted by the 1973 coup that installed Augusto Pinochet as the country's dictator. But it would take another infamous name to establish salmon in the country, this time as a farmed fish: Union Carbide, the chemical company known for the gas leak disaster that killed thousands in Bhopal, India, in 1984, was the first to build an open-ocean salmon farm in Chile.

After a slow start in the 1970s and '80s, the Chilean salmon industry became an enormous success—not that Chileans themselves were eating the fish. As soon as they got to market weight, the fish were shipped back to their native Northern Hemisphere. In the early years, Chile sold the majority of its salmon to Japan and the United States. By the 2000s, the industry had fairly exploded. It grew 25 percent a year from 2003 to 2006, producing 630,653 tons of salmon during that period, in an industry worth $2.335 billion by 2008. Chile became the world's second-largest producer of the fish, just behind Norway. By that point, the United States, especially Wal-Mart, was Chile's biggest customer. Nearly 80 percent of its salmon ended up on American plates.

But Chile's success, even as it skyrocketed, teetered on the brink of environmental disaster. Here's why: The salmon were raised using the same methodology as the worst land-based industrial feedlots. Salmon pens—circular nets submerged in Chile's fjords—were packed as densely as possible with the squirming fish. On average, the pens were filled with 42 fish per cubic yard. This created multiple environmental problems. First, the fish produced enormous amounts of waste that blanketed the seafloor under the pens. An average salmon farm with 3.5 million fish produces as much waste as a city of 169,000 people, creating a vast dead zone devoid of oxygen and life. And the salmon, exotic creatures in these fjords, escaped in huge numbers—an estimated 10 million during the peak production year—and wreaked havoc on the local ecosystem, including spreading disease and competing with native fish for food.

Speaking of diseases, the crowded conditions of the pens meant that infections were always possible. To stave off disease, the companies farming the salmon fed them massive amounts of prophylactic antibiotics, on the order of 300 times the amount used in Norway, where fewer fish are kept in pens that are also farther apart to reduce the risk of spreading infections. Some of the antibiotics aren't approved for use on livestock in the United States due to the risk of promoting antibiotic-resistant organisms that infect humans as well as fish. This included ciprofloxacin (Cipro), the powerful broad-spectrum antibiotic famously used to treat victims of anthrax attacks and one of the few that is often effective—as yet, anyway—when other antibiotics don't work. Farmed salmon are also susceptible to infestations of sea lice, which are parasites that attach themselves to fish and survive on a diet of mucus and skin cells. In the Northern Hemisphere, this causes a problem primarily for wild salmon, which can also contract them as they swim in the waters surrounding fish farms. For juvenile wild salmon passing through, sea lice picked up from a farm pen can be fatal.

By 2008, the environmental stress was too much for Chilean salmon farms. An outbreak of infectious salmon anemia (ISA) blazed through the pens, killing millions of the fish.

Although ISA doesn't affect people, it damaged Chile's fish-farming credibility. Major US retailers like Safeway and Wal-Mart stopped buying Chilean salmon. The virus, which decimated salmon farms in 2008 and 2009, forced the closure of dozens of farms and cost Chile $2 billion in revenue and 26,000 jobs. By the time it was contained, Chile was producing just half as much salmon as it did before the outbreak.

Even as ISA raged on, however, the industry was still trying to expand southward into Patagonia, Chile's treasure of untouched glacier-fed fjords. Oceana, which has a small and dedicated staff in Santiago, pushed back hard, as an industry that had already trashed coastal waters with disease and pollution had no standing to move on to pristine areas. Thankfully, by working with our allies, we've been successful so far, although the campaign is ongoing. Even more encouraging is that after the disastrous ISA outbreak, the Chilean government adopted most of Oceana's suggestions for ensuring the sensible management of the salmon business in Chile, including criminalizing fish escapes and mandating a series of measures to drastically reduce the use of antibiotics. This has forced the salmon companies to lower the density of the pens and alleviated many of the problems associated with overcrowding.

But even if Chile—and the other big farmed-salmon producers, like Norway and Canada—manage to stay virus and pollution free, it still won't solve the basic challenge of salmon farming our way to ocean abundance. There's one major sustainability problem with farming salmon: The fish is a carnivore. It eats *other wild fish* in the form of processed pellets of fish meal. So when you eat farmed salmon, you're eating wild fish all the same. Just how much fish it takes to fatten a salmon to market size varies around the world, and different companies make different claims. In the early years of salmon farming, it could take a staggering 10 pounds of wild fish to create 1 pound of salmon. The ratio has improved a little over the years; Stanford University scientist Rosamond Naylor pins it now at about 5 pounds.

Most of the fish meal is made up of small fish like sardine, anchovy (which is the largest fishery in the world by weight), and herring. While these are all perfectly edible and delicious fish, humans aren't their primary consumers. Instead, they're caught in huge quantities and ground into fish meal and fish oil. Eighty-one percent of the world's fish meal and fish oil is used for aquaculture. Scientists and conservationists call these "forage fish." They're low in the food web, reproduce fast, and perform a vital function in the oceans by feeding the dolphins, seabirds, and big fish that are the marquee species under the waves. In the industry, they have a different name that reveals how the business views these vital creatures: "reduction fish."

The huge number of forage fish ground up to feed farmed salmon, as well as the fish in smaller but growing aquaculture sectors such as cod and tuna, constitute a major stumbling block for people who want to see aquaculture as the solution for our depleted oceans. Even so-called organic salmon farms like some in Ireland and Scotland haven't quite figured out this puzzle.

Aquaculture should *add* edible protein to the world, not produce it at a net loss. Even if the industry's feed conversion ratio, as it's called, continues to improve, it still doesn't change the fact that we're farming more and more fish every year. The industry is expanding faster than the ratio can be reduced.

That's why we advise caution when basing your preferences for organic labels on farmed fish at this point, especially when they are carnivorous fish like salmon. While there may be a few cases where the "organic" fish on your plate meets your environmental expectations, in many cases it will not, primarily because many fish farmers are still feeding their carnivorous fish large amounts of fish meal, thus depleting the oceans. Some great carnivorous fish-farming operations have good feed-to-fish ratios—like the mullet that chef Dan Barber uses for a recipe later in this book—but, on the whole, they are not necessarily what you think you are getting when you order organic fish.

Fortunately, there are better choices. Not all aquaculture is created equal. Not every farmed marine creature is a voracious predator gobbling up resources as it is prepared for market. Some familiar options are the plant-eating tilapia, catfish, and even carp (if you're feeling adventurous)—which are sometimes "finished" with fish meal but are generally more ocean friendly. And then there is one of the world's most sustainable forms of protein that is also one of the healthiest foods: shellfish.

THE SOUTHERN TIP of the Delmarva Peninsula forms half the gateway separating the Atlantic Ocean from the Chesapeake Bay. Down here, just 25 miles as the crow flies from the 1.6 million people living in the sprawling suburbia of Hampton Roads and Virginia Beach, the beaches of Virginia's Eastern Shore are tranquil crescents of sand and dune grass. The nearest town, Cape Charles, is a cozy grid of turn-of-the-century homes on streets with names like Plum and Strawberry.

The Ballard Fish and Oyster Company facility just north of town doesn't look like much. It's merely a cluster of windowless warehouses overlooking the bay. But this facility, known as Cherrystone, can produce up to 100 million littleneck clams every year. The company produces every size of clam from chowder to button, but the littleneck is its most popular varietal, perfect for steaming and serving with oil and garlic. The names denote the sizes, which are measured across the hinge where the bivalve's two shells meet, but they're all the same species: *Mercenaria mercenaria*, a hard-shelled clam native to these temperate waters.

For centuries, the Chesapeake Bay produced millions of tons of wild clams and oysters. Many locals still believe the Algonquin name *Chesepioc* means "Great Shellfish Bay." (It doesn't. It was probably just the name of one of the Native American towns along the water's edge.) But the legend still speaks to the region's strong identification with water, watermen, and shellfish. Now the once-astonishing bounty extolled by Elizabethan explorer John Smith has become one of the most famous collapsed fisheries in the world. Overfishing, coupled with pollution from the growing cities on the Eastern Seaboard, habitat loss, and the proliferation of disease, has cut mollusk populations to a tiny fraction of their former size. The population of the eastern oyster, which famously crashed and is one of the most studied fishery collapses in America, has declined by 99.7 percent. Less attention has been paid to the Chesapeake's hard- and soft-shelled clam populations; but their numbers, too, have suffered dramatically.

Starting in 1895, the Ballard family watermen caught wild oysters in the bay, and by the middle of the following century, Ballard was one of the biggest names in oysters. But the family couldn't escape the drastic decline of the wild bivalves, and by the early 1980s, Carroll Ballard and his son Carroll Jr. were looking for other options. It took them 4 years of experimentation to grow the East Coast's first farmed clam. Now under the stewardship of Carroll Ballard III, who goes by Chad, the company is the world's largest producer of littleneck clams and one of the leaders in the still-nascent industry of farmed Virginia oysters. The company is well regarded for its environmental practices, even winning the title of Conservationist of the Year from the Chesapeake Bay Foundation in 1994. As Chad puts it, "We definitely see the need for sustainability in the next generation because we've been doing it for five of them."

Why does the farmed mollusk industry earn accolades while farmed salmon fends off bad press? It's the combination of a few factors and some luck. First is that there's no discernible taste difference between wild and farmed bivalves, which develop their flavor from the waters of the individual cove or inlet in which they're placed on floating racks or in net bags hitched to rebar on the bay floor to grow to market size. If anything, the more predictable shapes and sizes of farmed clams and oysters help their saleability. Farmed oysters are perfect on the half shell, as they grow individually into smooth cups rather than the irregular clumps found in the wild. And perhaps most famously, farming shellfish helps to heal the pollution-ravaged Chesapeake. In the 1880s, when historical records of their numbers put them at their peak, the oysters in the bay filtered and cleaned 19 trillion gallons of water—that's what the whole bay holds—in a matter of days. Now it takes more than a year. The farmed clams and oysters planted in the bay to grow to market size help replace the cleaning powers of their dwindling wild brethren.

The mere presence of shellfish, whether farmed or wild, also improves the immediate habitat. Robert Rheault, an oyster farmer in Rhode Island who's now executive director of the East Coast Shellfish Growers Association, described his patch of sea bottom as "black, anoxic mayonnaise" before he started setting oyster cages. Soon he had a vibrant ecosystem of fish, eelgrass, and lobsters surrounding the impromptu artificial reef. He says 1 ton of harvested oyster meat represents the removal of 4 pounds of nitrogen from the water, or the waste contribution of 38 people a year.

Clams, too, scrub the water with their gills as they feed. Rheault is experimenting with planting clams in New York City's Bronx River to see if the little mollusks can help improve the water quality there, too. "We can provide jobs, we can provide nutritious, delicious seafood, and have ecosystem benefits in terms of nitrogen, water quality, and habitat," Rheault says. "It's a win-win-win."

But the most important factor that makes farmed mollusks globally sustainable is the same one that is the undoing of farmed salmon: food. Clams and oysters feed at the lowest level of the food web, using their whipping microscopic cilia to pump water through their accordion-like gills and siphon out algae and other plankton.

At the Ballard Cherrystone facility, newborn clams are raised in tanks inside a hatchery that's constantly fed by an upwelling of Chesapeake Bay water. In a separate room, however, the staff supplements their diet, not with resource-intensive protein or some artificial Frankenfood, but with delicious house-raised algae. Festering in a row of Erlenmeyer flasks and varying in shades from neon green to a rusty brown, the different species of algae are specialized for different life stages of the mollusks. In the early stages of their lives, inside the hatchery, 150 million microscopic clams feast. Only about half will survive to the next phase, which takes place outdoors in another series of tanks flushed with water from the bay. Eventually, the surviving clams (each phase kills another 50 percent or so), now the size of nickels, are planted in grids in the shallow water just offshore, where they will grow for 2 to 3 years before workers rake them up and send them to the on-site processing plant for distribution around the country.

While raising the sensitive clams can have plenty of pitfalls that result in a low harvest, their life cycle on farms still hews closely to that of their wild kin. And once in the water, the shellfish valiantly perform their duties as water scrubbers. "What we're trying to do is what Mother Nature does, but at a much higher quantity," says Chad Ballard.

It seems that we should launch a consumer campaign to encourage people to eat more mollusks rather than the top predators we're in the habit of cooking for dinner. "It's our patriotic duty to eat as many farm-raised shellfish as we can," chef Barton Seaver has said, and he's on to something. But shellfish aquaculture does have some growth limitations. First is the difficulty of finding suitable undeveloped waters. The Ballard Cherrystone facility is mostly surrounded by land held by longtime private owners who don't mind watching the harvest from their windows. But a cluster of brand-new vacation homes in cheery pastel tones next door has the company on edge; newcomers might not be as appreciative of the region's historic ties to the livelihoods of watermen. There's also a biological limit to how many shellfish you can grow per square meter of seabed. Plant too many, and you risk starving the whole lot.

Even though aquaculture is the world's fastest-growing food sector, outfits like Ballard's are rarities because the United States simply doesn't have very many fish and shellfish farming operations. The number one farmed fish in the world is China's carp, the same fish that was the original farmed species thousands of years ago and still connotes good fortune and luck in China. Most of that carp stays in the country. But other farmed species enter the United States and Europe in a steady stream from the developing world: Indian shrimp, Vietnamese catfish, Mexican scallops. The global aquaculture market is worth

$100 billion, but the United States' contribution is just $1 billion. Since we're one of the world's largest importers, it adds up to a seafood trade deficit of more than $10 billion. And most of this seafood isn't inspected as it enters our borders, creating the potential for real health hazards like listeria, salmonella, and scombrotoxin infections. More than half of the United States' imported seafood comes from developing countries, many of which allow livestock and seafood to be treated with antibiotics and chemicals not permitted in this country. Even with these treatments, exporting countries aren't entirely successful in sending safe seafood. Of the imported seafood that was rejected at the US border between 2003 and 2006, more than a fifth was infected with salmonella.

So why not farm more fish domestically, where we can keep an eye on things? Most American farm-raised fish are Mississippi channel catfish, a sustainability success story because they are raised in self-contained freshwater tanks and fed a vegetarian diet. And our shellfish sector, as the Ballard outfit shows, is one of the cleanest methods of food production in the world. There's no technological barrier to growing more mollusks. In 2004, Colin Nash, veteran government fisheries scientist, called tripling our oyster, clam, mussel, and scallop production by 2015 "eminently possible." Ramping up production of finfish like salmon, haddock, and cod would prove more challenging because of the need for offshore infrastructure and scientific know-how on raising untested species, he wrote, but it was "technologically feasible given an immediate start."

Nearly a decade later, Nash's projections are largely unmet. The system we have now, in which cheaply raised foreign seafood is readily available, keeps most people pretty happy. And the wild capture industry reacts as though it's a threat. In Alaska, where wild salmon is king, salmon and many other forms of aquaculture have been outlawed. Conservation groups like Oceana are also concerned about open-water marine aquaculture because, shellfish aside, the industry is plagued with too many environmental and sustainability problems to trust our oceans with it.

Hauke Kite-Powell, a marine scientist at Woods Hole Oceanographic Institution, has studied the possibility of expanding aquaculture in the United

States, and even he doesn't think we're going to increase domestic fish farming anytime soon.

"We have a situation in the United States where we're wealthy enough to be able to import all the seafood we want from other places where attitudes are very different about this, and priorities are very different," he says. "So long as we place a higher priority in the use of our coastal waters on recreation and aesthetics than on food production, then it's going to be hard to see major growth in US aquaculture."

Still, aquaculture is exploding in dozens of other countries, from Norwegian cod to Thai shrimp. When it provides a net increase in the amount of edible protein, and does it at little environmental cost, aquaculture can be a great thing. When it provides premium protein at a net loss, however, it's only exacerbating the problems of our diminishing ocean resources and affordable seafood being taken away from needy populations. Better stewardship of the existing wild fish populations is not just a more environmentally friendly choice, it's also a better economic and humanitarian choice.

CHAPTER 6

The Fish We Don't Eat

No other of the finny tribe has so fine a flavour [as anchovies].

—MARY EATON

DESPITE GENERATIONS OF OVERFISHING, mismanagement, and exploitation, there are still parts of the oceans where giant schools of fish gather in epic numbers and provide millions upon millions of tons of healthy protein. There's just one problem: We're not eating them. Not directly, at least.

The world's largest fishery is found in the crisp, cold waters of the South Pacific, hugging the jagged coastline of South America. Here, schools of the slim, silver-scaled anchoveta, a small member of the anchovy family, are so enormous that the annual catch is three to five times the size of the world's second-largest fishery, the pollock of the North Pacific. At its peak in the 1960s, 1 in 10 pounds of the world's seafood catch was anchoveta. The fish are highly sensitive to El Niño, the meteorological phenomenon that warms coastal waters in the tropics and forces the schools to retreat to pockets of colder water. This makes the anchoveta even easier to catch, leading to all-too-predictable overfishing followed by years of recovery. Fishery managers in Peru and Chile have gotten a bit smarter about how to avoid taking juvenile fish, and in recent years, the anchoveta catch has stabilized, although it's still affected by severe El Niño events. Even at a reduced level—5.5 million to 8.8 million tons of fish caught each year—the anchoveta fishery is still bigger than the next three wild fisheries *combined*. But this is the world's biggest fishery not just because of its management, which has had its ups and downs over the years. Anchoveta are also resilient fish because they're small and reproduce quickly. When left alone, anchoveta schools are capable of doubling their size within 15 months. Compare that with cod, which can't reproduce until they're at least 6 years old, or the patient orange roughy, which isn't sexually mature until the ripe old age of 30.

But anchoveta aren't bound for sandwiches or dinner plates. Only about 2 percent of the catch is eaten directly by people. There are two reasons for this. First is the belief that anchoveta are inedible. Even our most fish-savvy scientists turned up their noses at the oily little fish.

"I worked in Peru in the '80s a lot," said Daniel Pauly, the fisheries scientist who initially uncovered the fact that global seafood catch was declining a decade ago. "At that time, it was agreed that anchoveta was not edible. I bought this because I was an idiot, and because everyone believed that."

Instead of being eaten by people, millions of tons of the anchoveta hauled up each year are pressed, dried, and milled into coarse brown powder. This is fish meal, and it, along with its by-product, fish oil, is exported around the world to be consumed by other animals: predator fish like salmon, pigs, chickens, and pet cats and dogs, even horses. A small percentage of fish oil ends up in nutritional supplements.

We already know that fish are incredibly healthy, omega-3-packed sources of protein. That's why livestock producers love to feed fish meal to their animals, even if the animal in question wouldn't know what to do with a fish if it jumped out of the water and into its jaws. But thanks to the huge demand from the livestock sector, anchoveta are big business—along with sardines, jack mackerels, and similar small, fast-growing marine fish that also get ground into fish meal and oil.

Yet all these fish are perfectly edible and delicious in their whole, unprocessed forms. Top chefs love fresh anchovies and other small fish (flip to page 131, and you'll find many wonderful recipes from many notable chefs that will let you discover just how tasty these small fry can be). These pungent fish "add an incredible depth of flavor" to a dish, according to sustainable-fish champion and chef Barton Seaver. When fresh (or freshly packed) and prepared simply and healthfully, these fish can supplant the shudder-inducing memory many of us have of our "fishy" first taste of canned anchovies or sardines. Oceana board member Ted Danson recalls, "The best fish I ever had was sardines in Basque country in Spain right off the boat, grilled with some olive oil and slapped between two pieces of bread." Hugh Fearnley-Whittingstall, the author and chef who promotes eating low-impact seafood with his Fish Fight campaign in the United Kingdom, adds, "Fresh sardines spitting and popping over charcoal create one of the most appetite-provoking smells I know."

In Spain, anchovies—or *boquerones*—as well as sardines and many other small fish species—are delicacies, not food for pigs. They're an excellent source of omega-3s, with more of the fatty acids per ounce than tuna, and their small size means they're less likely to have toxins in their flesh, which build up with age and size. Americans may not make a habit of eating these small, bony fish, but in southern Europe, Southeast Asia, and Africa, eating small fish is as common and as much a source of pleasure and gastronomic delight as eating tuna or grouper is in North America. They can be eaten whole, with the bones providing a third or more of an adult's recommended daily calcium intake, or filleted and served with salads, pasta dishes, or anything else you can dream up. Fearnley-Whittingstall adds, "They are delicious crisply fried, and crunchy enough to eat head, tail, and all—though equally easy to nibble off the backbone, leaving a classic pile of little cartoon fish skeletons on your plate!"

And yet forage fish like anchoveta and sardines are leaving the developing world by the millions of tons as fish meal and fish oil, destined to end up in the bellies of livestock and farmed salmon in America, Australia, China, and other wealthy nations. It's simple economics: Given the lack of appetite in much of the developed world for anchovy-based meals, the fish is most valuable when it's converted into other forms of animal protein via the guts of pigs or cows or chickens. This means that basic market-driven dynamics of the world are reducing the available protein for the nations that need it the most.

In Peru, where vast schools of anchoveta produce the majority of the world's fish meal, nearly a third of the population lives below the poverty line. And the hulking processing plants where the fish are ground up create a tragic dichotomy, with the people unfortunate enough to live nearby the losers: As the food product helping to fatten fish dinners in wealthy cities like New York and Dubai churns inside those factories, local Peruvians struggle with the pollution they emit. Billowing smoke forces people to stay in their homes and causes widespread asthma, and water and food are contaminated with effluents that even played a role in a cholera epidemic.

In the coastal town of Chimbote, life expectancy is 10 years less than the national average thanks in part to the smoke-belching fish meal plants. Maria Elena Foronda Farro is a sociologist who grew up there just as the plants were cropping up in the 1960s. For years, she has worked to get the processing plants to use cleaner technology. Foronda was repaid for her public health activism with a year in prison after she and her husband were falsely accused of belonging to a terrorist group in 1994. But she remains undeterred. Since leaving prison, she's convinced numerous plants to switch to cleaner technology, and she told *Grist Magazine* in 2003, "It's my life's work. . . . You only need a tiny spark of social consciousness to become an activist."

Foronda is one of a small group in Peru who are nobly working to reduce the impacts of the processing plants on the communities that surround them. But the larger impact of these plants is that they keep perfectly good, cheap protein away from the people who need it the most. Oceana's scientists evaluated the amount of fish that ends up in the processing plants and found that if you took all the world's fish meal–bound fish and instead fed it directly to people, you could provide an additional 400 million healthy fish dinners a day.

Peruvian chef Gastón Acurio, who runs a restaurant empire throughout South America, has helped spearhead the effort to remove the stigma of anchoveta as an inedible fish by featuring it on the menus of his restaurants. Right now virtually all the anchoveta is exported from Peru because there's no local market in place to provide demand and no place to buy or eat anchoveta. Acurio is demonstrating that this fast-reproducing, zooplankton-eating fish deserves to be reconsidered. His efforts have shown some success already: Direct consumption of anchoveta in Peru has increased from 11,000 tons in 2006 to 210,000 tons in 2010. That still represents 3 percent or less of all the anchoveta pulled from the water, and so Acurio's campaign continues with the help of Peru's former deputy minister for fisheries, the marine scientist Patricia Majluf. She met with fierce opposition when she called out fishermen for selling anchoveta they'd been licensed to catch for human consumption to fish meal plants, where they fetch a higher price. Between the two of them, the march to wrest anchoveta from the lucrative fish meal industry for the benefit of the poor has begun. Obviously, it's going to be an uphill climb.

FORAGE FISH LIKE ANCHOVETA are usually small and reproduce quickly, which means they're potentially highly productive, highly sustainable fisheries. That doesn't mean we've always been good stewards, however. In Chile, an 18-inch-long blue-gray forage fish called the jack mackerel (*jurel* in Spanish) was once the fourth-largest fishery in the world. In the 1980s, at its peak, Chileans caught 5 million tons of the fish every year. Alex Muñoz, who leads Oceana's campaigns in Chile, remembers seeing cans of *jurel* at the local market as a child. "It was used to feed pets," he said. "Now you can hardly find it. Nobody would feed a cat jack mackerel today."

That's because the Chilean fishing industry catches only about 198,000 tons of the fish these days, 4 percent of its peak just a couple of decades ago—another casualty of greedy overfishing. Muñoz and his staff have succeeded in getting the government to set quotas based on scientific recommendations it had previously ignored, but it may be too late for the jack mackerel to rebound to its once-incredible bounty. Time will tell. It's too bad—we hear that jurel, like mackerel, can be a wonderful, healthy meal. It is an oily, strongly flavored fish (like mackerel) and is often baked in a savory sauce or broiled (sometimes with a citrus marinade to tone down the fishiness).

The highly profitable fish meal industry is increasingly squeezing out the local artisanal fishers. In Chile, for example, only 5 percent of the jack mackerel quota is given to artisanal fishermen even though they outnumber the industrial fishers by thousands. This is an enormous, and so far missed, opportunity. Local fishers must operate within 5 miles of the country's coastline, where forage fish abound. It is these fishermen who will be able to provide anchoveta, sardine, and jack mackerel for direct human consumption in Chile and Peru. But with only a tiny percentage of the catch afforded to these fishermen and little infrastructure like processing plants and markets in place to support human consumption of these fish, they'll continue to mainly be ground into fish meal and exported to wealthier nations, bypassing the local population.

One-third of all the wild fish caught on Earth end up as fish meal or oil. Of that, 81 percent goes to feed farmed fish. Given how rapidly the aquaculture industry is expanding, it's clear that the demand for forage fish won't abate anytime soon. But the market may finally begin to play a helpful role; increasingly expensive thanks to the price of fuel, forage fish are getting more and more pricey. The aquaculture industry, as well as livestock producers, are motivated to replace fish meal with something else.

But that day may not come soon enough for the oceans and for the people who could be enjoying nutritious seafood meals today. The first step is to remove the stigma of eating what many perceive as inedible fish. As Daniel Pauly told us, "They must be treated how you treat food." The same might go for the menhaden found all along the US East Coast, one of the most popular species for reducing into fish oil and widely considered inedible because of its oily, bony, and incredibly fishy nature. "I have always read that [menhaden] is not edible by people," Pauly continued. "And I wonder if it is true. I have never heard of anyone taking a fresh one and actually working on making it a human product. It's bony, but boniness is in the eye of the beholder. Cod are bony and are the preferred fish of millions of people." A few brave fishers are already eating menhaden—more commonly known as bunkers, pogies, mossbacks, bugmouths, or fatbacks—and enjoying them barbecued (like bluefish) and stir-fried.

Chef Alton Brown has aligned himself with the Sardinistas, a small cabal of culinary-minded artists, writers, and fishers in California who aim to bring sardines back to the American palate. In the first half of the 20th century, California's sardine industry was so vast that 700,00 tons of the fish were canned every year on Monterey Bay's Cannery Row. After World War II, the fishery collapsed under the strain of decades of overfishing, and now the oceanfront streets of Cannery Row are filled with trendy restaurants and shops.

Brown loves sardines so much that he takes a tin to eat with chopsticks for lunch each day when he's on the road filming *Good Eats*. But even he, the plain-spoken purveyor of consumer-friendly meals, recognizes the challenge of getting a little fish like the sardine back on dinner plates. "Americans: We're people of the cut, not the carcass," he told the *Washington Post*. "We need to teach our children that, yes, it had a face. And, yes, it had a life. And here, it has fins." And then perhaps the idea of eating a whole fish rather than a single fillet would not be so daunting.

And forage fish can be more than just tasty. They provide jobs. They reduce pollution. It seems counterintuitive to argue that eating fish can add to the world's available protein, but if you eat anchovies instead of farmed predator fish like salmon, that's exactly what you're doing. This speaks to what scientists call the fish-in, fish-out ratio. As we saw in Chapter 5, it takes up to 5 pounds of wild forage fish to create 1 pound of farmed salmon—resulting in a net loss of protein. If you took the "reduction" fish—the anchovy, the sardine, the jurel—that was fed to the salmon and instead made *it* your dinner, you'd really be eating one fillet of fish when you eat one fillet of fish, not five fish that aren't there. You'd be leaving more fish in the ocean.

And we haven't forgotten the dolphins, whales, seabirds, sharks, and more that rely on forage fish to survive. Giant schools of these fish form an intermediate step in the food web: They eat microscopic plants, and the big predators eat them. An ocean devoid of forage fish would be an ocean without these signature creatures. So our directly consuming these little fish, funny as it may seem, could help protect the entire marine food web. In a world where we are scandalized by schoolchildren eating "pink slime," the leftover trimmings of industrial beef production, there are few things more natural than a whole fish provided by Mother Nature, whether it's bony or not. And there's no industrial process—no additive, no extraction, no modification, no passing through the intestines of a feedlot pig—that can improve upon the original creature, the perfect protein: the fish.

CHAPTER 7

The Terrestrial Trap

As for diversity, what remains of our native fauna
and flora remains only because agriculture has not
got around to destroying it.

—ALDO LEOPOLD

LIFE ON EARTH began in the ocean, but conservation began on land. Whether you date its roots to the Sierra Nevada tramps of naturalist John Muir, the wilderness studies of ecologist Aldo Leopold, the pond gazes of Henry David Thoreau, the national parks of Theodore Roosevelt, or Rachel Carson's forecast of a silent spring, the "green" agenda developed on land for decades before anyone seriously considered initiating its blue counterpart.

This isn't surprising. While oceans cover 71 percent of our planet, humanity inhabits the other 29 percent, and all politics is local. We've maligned Thomas "Inexhaustible Sea" Huxley, but he was a perceptive and courageous thinker who earned one of history's gold stars by rising early to the defense of Charles Darwin. Like nearly everyone, he just had a blind spot when it came to the oceans. Grasping the immensity of the seas, he couldn't imagine that humanity could ever present a fundamental challenge to life beneath the waves.

Meanwhile, on land, you could watch as centuries-old forests fell before advancing Pilgrims, milled into sawdust and plowed into cornfields. In 1872, the United States established its first national park, Yellowstone. But more than 100 years passed before the creation of the first comprehensive national marine fisheries law. Before the passage of the Magnuson-Stevens Fishery Conservation and Management Act in 1976, "federal management of marine fisheries was virtually non-existent. With the exception of state managed waters, federal activities were limited to supporting a patchwork of fishery-specific treaties governing international waters, which at that time existed only 12 miles off our nation's coasts," explains Eric Schwaab, formerly the top administrator at the National Marine Fisheries Service of the United States. So marine conservation policy making is young. Its progenitors—many still well short of retirement—framed the task of saving the oceans in ways rooted in the long-standing experience of terrestrial conservation.

On the land, agriculture is at odds with biodiversity. We cut down and plowed up the biodiverse places—the forests and prairies—to plant the cornfields. Over and over again, in farms and fields all over the world, maximizing food production for people directly conflicts with maximizing the opportunities for wild creatures.

As a 2007 study in the journal *Science* points out, "Agriculture has been the major driver of biodiversity loss in many ecosystems." But this logic is a tough sell with people who need food. "We cannot feed our children because the government wouldn't allow us to till the forest land; who cares for some environment that doesn't help us?" complained Virginia Wanjiku, a 58-year-old Kenyan vegetable seller, to Sam Aola Ooko, a journalist surveying locals about their views on environmentalism.

Take Madagascar. This island nation in the Indian Ocean is one of the world's wildlife meccas. Its isolated rain forests are home to dozens of species of lemurs and hundreds of species of frogs found nowhere else on Earth. Just 20 percent of Madagascar's rain forest still exists and it is one of the world's poorest countries. Four out of five families depend on subsistence agriculture to survive. Would you tell a poor Malagasy farmer he can't take that extra acre of land to feed his family?

Here in the United States, we don't face acute hunger crises like some developing nations. Our battle for biodiversity tends to pit conservation against economic development. For years, the northern spotted owl was the symbol of this conflict. Six thousand of these little-seen birds were living on millions of acres of federally owned old-growth forest in the Pacific Northwest, and they lit up a national debate when the owl was protected under the Endangered Species Act in 1990. The timber industry claimed it would lose tens of thousands of jobs, inspiring bumper stickers with slogans like "Save a Logger, Eat an Owl." More than a thousand timber-town residents rallied, chanting "Families first, owls last." Twenty-two years later, both the owls and the logging industry are struggling to survive. And for many people, "spotted owl" is still shorthand for government decisions that they think unreasonably favor biodiversity over industry and, ultimately, people.

Many years of "spotted owl" conflicts on land have informed the way we now look at saving the oceans. The basis of the save-the-forest debate is that humans can't both optimize the land's productivity by logging it—or turning it into cropland—and save the wild animals that live there. Conservationists therefore prioritized protecting biodiversity over producing food. We have assumed it's the same in the ocean.

So as conservationists ventured into the salt water, we made biodiversity our agenda. One illustration of this is the Marine Mammal Protection Act of 1972, which passed 4 years before the Magnuson-Stevens Act first provided even basic rules for national fishery management. The act's purpose is to protect marine mammals from being killed or injured by commercial fishing operations. It explicitly required the big commercial fleets to change how they fished in order to protect animals that are not caught for food in the United States. Its restrictions on the use of certain gear and the time or place of fishing were perceived by some in the fleet as a hindrance to the efficient catching of food fish, and still are today.

Saving the whales and the other creatures that are protected by the act is one of the signal achievements of the conservation movement in the United

States. And Oceana has campaigned in recent years to beat back proposed changes that would weaken its protections. But let's not overstate how much it's meant to the overall health of the oceans. Even with the subsequent enactment of a basic fishery management law (Magnuson-Stevens) and its strengthening under President Bill Clinton, we still have important work to do to achieve an abundant and fully productive ocean. In 2012, 44 American fish stocks were considered overfished by the federal government. Populations of some fish in New England and the Northeast have dipped so low that the government issued a federal disaster declaration that opened up the possibility of disaster funds for fishermen.

Ocean conservationists' preoccupation with biodiversity is also evident in the choices big international conservation organizations and philanthropists make about where in the world to work on ocean conservation. They have tended to focus on the most biodiverse areas of the world's oceans. These are areas with large numbers of different species per square mile of water or with many species unique to that place. A recent study showed that these areas have seen the greatest amounts of philanthropic investment per ton of fish.

The most biodiverse areas of the ocean tend to be hot, reefy places. But the most *productive* areas (with a few notable exceptions like Indonesia), the places where we find huge volumes of schooling marine fish, tend to be cold or temperate places. As a result, conservationists' focus on saving biodiversity has meant that these areas of the world's oceans have been underserved. Per ton of fish, ocean philanthropists as a group are investing less than one-quarter of their dollars in productive areas, as opposed to biodiverse ones.

This investment choice is especially troubling because the world's ocean fisheries are in trouble. Despite big advances that have been made in fishing technology, the total catch of the world's commercial fishing fleets peaked in the late 1980s and is now in decline. The fish aren't there to be caught.

We have been failing to pay real attention to the management of the most productive parts of the ocean. That leaves us with a huge opportunity.

This opportunity may seem counterintuitive. Our experience on land has taught us that biodiversity and productivity usually go hand in hand. The areas on land with the greatest diversity of creatures tend to be where life is most abundant. Just think of rain forests: The world's tropical forests contain 80 tons of life per acre—almost double what you would find in a temperate North American forest. And they are home to an outsize share of the world's species: Roughly 10 percent of all the world's species can be found in the Amazon Basin, which accounts for only 4 percent of the world's land mass.

Under the waves, the fundamental driver of productivity—the tons of marine creatures that grow in a cubic mile of the surface ocean water layer per year—is determined by the rate of growth of the creatures at the lowest level of the food web. An abundance of organisms at the base of the food chain means the creatures that eat them will also be abundant, and so on, up through the five trophic levels that scientists use to order life in the sea. Algae and other photosynthetic forms of life are at the bottom, called trophic level one. Since they live off the energy of sunlight, you might guess that the most abundant areas of the ocean would be those with the most sunshine—the tropics. But it's not that simple. The northern and southern latitudes have plenty of sunshine, too. The limiting factor in ocean photosynthesis isn't sunlight. Instead, it's the key mineral nutrients that feed these microscopic ocean plants.

As a result, the most productive areas of the world's oceans are typically the ones where these mineral nutrients are most generously and consistently available at the surface of the ocean, where the sunlight also penetrates. There are two natural sources: upwelling currents from deep, mineral-laden zones of the ocean and wind-carried mineral dust from dry coastal areas. Of the two, the first is the most important in powering the oceans' ability to host a cornucopia of life. Consider the country with the most productive ocean in the entire world: Peru is all by itself the source of 11 percent of the entire world's wild fish catch by weight.

Why is this? The Peruvian ocean waters consist of the ideal conditions to support enormous masses of fish. The molecules of cold water take up less space than warmer waters, so in most parts of the oceans, cold water sinks while warm water sits on the surface, warming even more under the sun's gaze. But in certain parts of the world, a perfect confluence of geography, currents, and wind allows the deep, cold water to be pushed upward to the surface in what's called an *upwelling*. The coast of Peru is one of these places. Crisp Antarctic waters shuttle northward along the South American coast on an international oceanic highway called the Humboldt Current. This brings the cold, mineral-rich waters to the surface in a food bonanza for zooplankton and phytoplankton, and likewise fish. It's what makes it possible for Peru to have such an enormous fishery. Peruvian anchoveta is the world's largest fishery, bigger than the world's next three fisheries combined.

There are five major upwelling spots in the world: Peru, California, northwest Africa, southwest Africa, and Somalia. These areas cover just 5 percent of the ocean but produce 25 percent of the world's wild fish catch.

These fish-filled places need our attention now.

Happily, in 1996, the Magnuson-Stevens Act was amended after a long debate in Congress. For the first time, the goals of optimizing fishery productivity and ecological system protection were given equal emphasis. You could hear it even in the floor speech delivered by Senator Jesse Helms of North Carolina, a conservative Republican and no friend of the environmental movement, just before he voted on the act: "Mr. President, I greatly enjoy seafood. I have dined in many seafood restaurants in coastal North Carolina and many fish houses further inland. North Carolinians want to maintain a steady supply of good, high-quality seafood well into the future. We can do that if our fishery resources are well managed in an environmentally responsible manner."

The law formally recognizes that, in the ocean, the warring forces of conservation and food production can be allies. Because we are catching and eating a wild resource produced by an ecosystem, the policies that optimize ecosystem health also optimize fishery abundance. Protecting nature in the ocean and managing fishery productivity are the same thing.

This win-win situation is so surprising to some that it continues to be strongly challenged. The biggest uproar comes from people who assert that since top predators like dolphins, sharks, and sea lions eat a lot of fish, optimizing the oceans' food potential for humanity requires eliminating the competition

of these creatures. It's true that they do eat a lot. An adult sea lion eats as much as 40 pounds of fish a day. Before the Marine Mammal Protection Act was passed in 1972, there were 5,000 California sea lions; today, the number is in the hundreds of thousands. That's a lot of sea lions eating a lot of fish. Some people argue that the sea lions are eating so much fish that we need to get rid of them. Sean Carpenter, a river guide for fishing trips in Oregon, told one reporter in 2006 that "there have been days where maybe a hundred fish were caught, hooked all day, and twenty-five landed. Seventy-five go to the sea lions. . . . I don't know how people are going to react to hunting them. But that's what really needs to be done." The federal government recently authorized the limited killing of California sea lions to, they claim, protect salmon.

The idea that we should eliminate predators that are our competitors is a bad one. Complex ecosystems are more resilient than simple ones. The diversity of creatures in a complex ocean system, and the multitude of intersecting roles played by all the species resident in an abundant ocean, mean that threats to one creature do not create cascading effects that bring down the stability of the entire ocean ecosystem. The resilience you get from this complexity and diversity is essential to the productivity of the world's oceans and their ability to give us lots of different kinds of food.

Here's an example. One of the longest-running annual surveys of shark populations is conducted off the coast of the Carolinas by the University of North Carolina Institute of Marine Sciences. In a paper published in the journal *Science* in 2007, researchers found that populations of 11 species of big sharks like bull sharks and hammerheads were, around the turn of the century, at single-digit percentages of their levels in the 1970s. By 2004, the commercial scallop fishery in North Carolina collapsed. The century-old fishery was closed due to the extremely low populations of bay scallops. Scientists determined that overfishing of big sharks caused the scallop collapse. Populations of scallop predators, especially rays and other small sharks, had boomed because their predators, the really big sharks, had been nearly eliminated.

It does not require a conservation agenda to seek protections for biodiversity in the ocean. A concern with food security will lead you to it. Like a

diversified investment portfolio, a highly diverse ocean is a more stable ocean. It is more resilient in the face of sudden changes. Over the long run, this stability makes it more productive. And the "long run" is not just a vague point in the distant future. This robustness is especially important right now, when the oceans face extraordinary challenges from global climate change.

Simplifying the natural food web by eliminating categories of creatures—for example, the big predators—is also an argument whose logic drives its own demise. Suppose you choose to eliminate all the sea lions, sharks, dolphins, swordfish, tuna, and other animals at the top of the food web in exchange for delivering the fish they've been eating to the plates of hungry people around the world. Consider where this policy would take you. As soon as these top predators—scientists call them trophic level five—are eliminated, there will be a new group of top predators in the ocean—trophic level four—consuming what seem like vast quantities of fish that could be fed to people. So by the logic that drove the elimination of the old group of top predators, these new ones will also be fished out. This will then produce a new set of top predators, and so on. In the end, we will be eating jellyfish and zooplankton. Or marine bacteria.

Some argue against the notion that a better-managed ocean would allow for increased fish catches. They think the human appetite for fish presents a fundamental threat. Offended by the fact that many of the conservationists he knew ate meat and fish, Sea Shepherd Conservation Society leader Paul Watson (of *Whale Wars* fame) wrote, "The problem is that people like Carl Pope, the executive director of the Sierra Club, or the heads of Greenpeace, World Wildlife Fund, Conservation International, and many other big groups just refuse to accept that their eating habits may be just as much a part of the problem as all those things they are trying to oppose."

Our suggestion that better-managed oceans can produce, on a sustainable level, more fish for humanity will generate angry calls of betrayal from some of our allies in conservation.

Take a deep breath, please.

We agree that short-term reductions in fishing are necessary to rebuild spawning populations. The ocean's bank account will generate more interest,

but only if we rebuild its principal. Yet many oceanic natural systems are so fertile that these reductions can, in just 5 or 10 years, produce increases in abundance that will support higher and sustainable catch levels. This is not clear only in theory. It is apparent in fishery data around the world. It just requires three commonsense principles of good ocean management:

1. Protect the habitats that foster ocean life.
2. Reduce the scourge of bycatch.
3. Set quotas based on science, not the fishing industry's bottom line.

This is not a pie-in-the-sky dream of saving the oceans. People around the world who have taken on these tenets have seen once-scarce fish come back to abundance. Let's explore a couple of their stories.

CHAPTER 8

Swimming Upriver

One man killed a Small Sammon, and the Indians gave me
another which afforded us a Sleight brackfast. Those Pore people
are here depending on what fish they can catch, without
anything else to depend on.

—WILLIAM CLARK, *ORIGINAL JOURNALS OF THE LEWIS AND CLARK EXPEDITION*

PERHAPS NO FISH on the planet has been the subject of more speculation and conservation efforts than the wild salmon. The history of this chameleonic fish will sound familiar by this point: Once so plentiful that the fish's return runs upriver—sometimes hundreds of miles—to spawn at the places of their births could be considered one of nature's greatest migrations, many salmon populations on the western coast of North America are now close to extinction. The East Coast used to have a healthy salmon population, too, but not anymore. Any "Atlantic" salmon you buy at the store is farmed, its wild brethren lost to history.

We haven't made it easy on Pacific salmon, either. We started with logging, sending massive tree trunks down the western rivers, gouging the beds the fish relied on to spawn and making the water too murky. We dumped leftover sawdust into the water. We dug out reservoirs and allowed toxic runoff from mines to seep into the watershed. In one famous disaster on the Fraser River in

British Columbia, the Canadian Northern Railroad Company blasted tons of rock into a narrow channel that had currents so strong it was already called Hell's Gate. The increased turbulence killed millions of sockeye, whose bright red carcasses blanketed the waters for 10 miles. That was in 1912, and the salmon didn't recover for decades.

In the New Deal era, we began erecting dams. The Bonneville and Grand Coulee dams were just the first of dozens of sheer concrete walls that would block the fish's ancient migration routes along the Columbia, the Pacific Northwest's greatest river. The engineers tried to compensate by building fish ladders into the sides of the monoliths, with curves so graceful they were featured in colored-pencil postcards. Some fish do still make it up the "fishways," as they're called, but this hasn't stopped the inexorable decline of the West Coast salmon that frustrates fishermen today.

It didn't help that this onslaught on the riverine habitat was paired with enormous rates of overfishing. In the early years of the 20th century, quaint rowboats with butterfly sails were rapidly replaced by gas-powered engines and trawl lines a hundred hooks long in the Pacific Northwest's Columbia River, at the time the world's greatest producer of coho and chinook salmon. Native Americans who had fished salmon sustainably for centuries were pushed out of their traditional fishing grounds, which, coincidentally, were also the best places to fish. They watched from the sidelines as settlers replaced their dip nets with 100-foot gill nets, traps, fish wheels, and more. In some years, the new fishermen took as much as 90 percent of the salmon biomass from the river.

Meanwhile, the canneries operated with brutal efficiency. In 1877, 300 workers packed up to 450 cans of salmon a day; just 5 years later, half as many people packed twice as much. It was the same as in any modern industrial fishery: few workers, huge impact. The annual chinook salmon production of the Columbia River peaked at 43 million pounds, an enormous, practically inconceivable amount, in 1883. By the end of the decade, chinook populations were unable to meet the industry's insatiable demand. Canners started targeting smaller coho and chum salmon, which are still sometimes called "dog" salmon because they were fed to, you guessed it, dogs. Canned salmon was a popular, easily stored, and cheap food throughout World War I. But as salmon runs declined, the price of the fish went up. Those diverging lines meant that fewer and fewer salmon were canned over the passing years. By the 1970s, the cannery business was all but dead. In 1980, the last Columbia River cannery closed, and with it, a productive river moved into the realm of memory. The intensity of fishing pressure and habitat destruction visited upon salmon populations in the western United States and Canada has been rivaled only by the slew of costly, and sometimes problematic, attempts to save the fish. We have spent billions of federal dollars to bolster salmon populations at the same time we've been methodically destroying their river homes.

Perhaps the biggest controversy in the history of salmon management was the exclusive focus on hatchery programs, starting in the 1870s when the first US fisheries commissioner, Spencer Baird, chose to pursue breeding more fish over protecting habitats or reducing overfishing. Early hatchery managers scoured riverbeds for salmon roe, removing genetic diversity from the ecosystem, and released the fry in rivers across the country without regard for the fact that each batch of young fish was genetically attuned to the specific river bend where its parents had spawned. The managers also largely ignored the reality that a salmon population's years in the ocean and the conditions there play an enormous role in the success of a salmon run. Pollution, water temperature, and population fluctuations—both natural and unnatural—in the salmon's prey of tiny crustaceans can create a historic boom of salmon cruising back to spawn, or a disaster of epic proportions.

Fisheries biologist Jim Lichatowich has documented the legacy of the hatchery program in one of his books, writing, "We assumed we could control the biological productivity of salmon and 'improve' upon natural processes that we didn't even try to understand. We assumed we could have salmon without rivers."

In Canada, on the Fraser River, habitat-protecting policies and fishing regulations have helped the salmon to return to multimillion-strong runs in recent years. It's a different story in Washington, Oregon, and California, where most West Coast salmon are in dire straits. On the Columbia, the 2011 chinook run numbered about 600,000, a shadow of the days when 10 million fish swarmed the river. Salmon are now extinct in nearly half of the rivers where they once spawned in Oregon, Washington, California, and Idaho. In 2008 and 2009, California's entire salmon fishery was shuttered for the first time in its history. The closure cost the state $250 million a year. There simply weren't any fish. A century of overharvest and river degradation had done its damage.

In his book *Salmon without Rivers*, Lichatowich recounts a Tlingit myth about young boys who threw live salmon on a fire. The spirits of the fish retaliated by killing the boys and other members of the village until the tribe promised to never again treat animals with disrespect. Lichatowich compares that to the canneries, where unused salmon carcasses rotted in piles: "This kind of overharvest and waste could never happen in a culture where people engaged in a personal relationship with the salmon, in a culture where people regarded the salmon as an important gift that had to be treated with respect."

Sadly, we still fail to heed the lesson of the story of the Tlingit boys. As valuable as the chinook salmon has become, both culturally and economically, there is still one place where the fish are treated like trash. In Alaska's Bering Sea, where salmon go to live out their adult years before returning to rivers to spawn, big factory trawlers have killed hundreds of thousands of the fish. Not to be sold, not to be eaten. These chinook salmon, the lifeblood of the region for thousands of years, are the bycatch of the just 3-decade-old commercial pollock fishery. And the multiyear effort by conservationists and Native organizations to control this bycatch illustrates one of the three tenets of our guide to smart fishing. Our approach pays dividends rather than eating away at the principal and leaving collapsed ecosystems behind: Protect habitat, reduce bycatch, and set quotas based on science. Do these things, and your fisheries can replenish plentifully every year. Don't do them, and fisheries like Alaska's salmon face the same bleak future that's already a reality in many of the rivers of Washington, Oregon, and California.

TWO GREAT RIVERS zigzag to the Bering Sea from the mountains across Alaska's western tundra. The Yukon and the Kuskokwim snake for thousands of switchbacked miles across a permafrost where a million tiny landlocked lakes appear in summer when the ice melts. This watershed is one of the world's largest. In addition to caribou, moose, wolverines, grouse, ptarmigans, and mosquitoes, the Yukon–Kuskokwim (Y–K) Delta is home to all five North American species of Pacific salmon: coho, sockeye, pink, chum, and the largest—the king—chinook.

Twenty-five thousand people, mostly Yup'iks, live scattered across the treeless plains, and their connection to salmon goes back to their beginnings: The Yup'ik word for fish, *neqa,* also means "food."

Myron Naneng's office overlooks the Kuskokwim as it passes through Bethel, a town of 6,000 that serves as the hub for dozens of Yup'ik villages in this region 100 miles from the coast. Bethel feels much larger than it is. It's home to the region's only hospital and penitentiary. In winter, the town hosts a 300-mile dogsled race. The summer is filled with fishing.

Naneng is aggrieved. He is the president of the Association of Village Council Presidents, a nonprofit that represents the Yup'iks of the Y–K Delta. Now, several dozen of his constituents are facing $500 fines for fishing chinook from the Kuskokwim after the state government decided on an early closure for the spring run when the fish's numbers failed to meet even low expectations. Chinook, which once supported a commercial fishery worth $10 million in the Y–K Delta, have been declining for years. The commercial fishery is gone. All that's left are the subsistence fishermen, whose right to fish salmon is supposed to be protected under Alaska's constitution.

Naneng, leaning back in his chair and knitting his hands across his belly, recounted a conversation he'd had 6 weeks earlier with Alaska governor Sean Parnell.

"I told him, 'Governor, if someone asks me, "Should I go fishing today, I need to go out fishing for food," what do you expect me to tell him?'" Naneng said, his voice quiet and firm. "He didn't respond, so I said to him, 'If anybody calls me and says, "Can I go out fishing today," I will say go ahead and do it. "Nobody else is going to provide it for you."' And what did he do? [He said], 'Oh, I'll wait for a report from my staff from the fish biologist and all that before we'll make a determination.' He could have called them right there, the director of Fish and Game, open it so people can go out for food. And guess what happened. People went out for food and they got cited."

During the closure, people were allowed to fish, but only with 6-inch mesh nets, which may catch smaller fish but allow the larger chinook to bounce out. Most people in villages and fish camps don't have easy access to new equipment. At more than $1,000 for a 300-foot net with a float and lead line, new gear is prohibitively expensive for Yup'iks living off the water. Many of them live a truly subsistence lifestyle and have little opportunity for getting outside income. They spend the spring and fall salmon runs fishing from campsites passed down through generations. These clusters of wooden huts and fish-drying racks line the Kuskokwim's brown expanse or are tucked away along the thousands of willow- and cottonwood-lined tributaries and sloughs edging the big river.

A family might need about 120 chinook to make it through the winter. It's backbreaking work to prepare the fish so it will last through the season. Every member of the family is involved, from setting and hauling in the nets to gutting and preparing the fish for drying on cottonwood racks for several weeks in May and June before the rains come in midsummer. Another run of chinook and chum comes in the fall. The fish, smoked and dried, are eaten with seal oil over the winter. In remote places like Bethel, where a gallon of gas costs $6.70 and a 10-pound bag of potatoes nearly $20, subsistence-sourced fish are still the cheapest and healthiest meals available.

A salmon dinner also keeps alive the Yup'ik traditional lifestyle. As Naneng said, "I'd rather have dried salmon for lunch than anything else. You appreciate your hard labor and work you've done to put that fish away and share with family."

Alaska's harsh climate and remoteness have helped stave off development like the huge dams that choked off rivers in the Pacific Northwest, knocking out thousands of miles of salmon habitat in single, poured-concrete strokes. The mining and logging industries that changed the landscape of the Lower 48 are well entrenched in Alaska, but fishing has long been prioritized and protected by the government and people. And yet chinook salmon have been in steady decline for the last several years. Naneng has been watching carefully. He has heard stories from villagers who work on trawlers and have seen the nets pull in 4-foot adult chinooks.

"I know that people in the river system are part of the problem," he said. "But I think the bigger issue is that out in the Bering Sea, they have a legalized wanton waste program. If you are a trawler, it's okay for you, because of the fact that the salmon are labeled a prohibited species. If you catch them, you have to throw them overboard whether they're dead or alive."

That the pollock fleet catches chinook salmon is no secret. From 1997 to 2006, nearly 50,000 chinook on average were caught by the Bering Sea groundfish fishery each year, and 90 percent of those were taken by pollock nets. This may not seem like that many fish, especially when you consider that the pollock fleet hauls in more than 1 million tons of their quarry every year. (Salmon are so valuable that they are the only species counted by the individual rather than the pound.) Consider this, however: The entire Native subsistence community of Alaska, all 25,000 households, harvests about 160,000 chinook salmon per year.

The pollock industry's trash is food, security, and life to these Alaskans.

As if taking 50,000 chinook a year wasn't bad enough, the catch suddenly spiked toward the end of the 2000s. In 2005, pollock nets snared nearly 75,000 chinook. In 2006, it was 87,771. In 2007, it reached a gut-punching all-time high: 130,000 chinook salmon dead in pollock nets before they could return to rivers in not just Alaska but also Canada and the Pacific Northwest. The worst part was that this was all legal. There was no restriction on how many chinook salmon the pollock industry could take.

Let's stop here for a moment to consider the pollock and the mammoth industry it supports.

TO BE BLUNT, the walleye pollock is no regal chinook salmon. This modest-size groundfish is a member of the cod family and gets its name from the white circles around its eyes that give it a perpetually surprised expression. Pollock travel in enormous schools. The juveniles feast on the cold North Pacific's copepods and krill, which are tiny crustaceans. The adult pollock feast, cannibalistically, on the juveniles along with other fish and krill. But the fish is so fecund that each female produces up to 2 million eggs over a couple-week-long spawning period, and so it survives even this insider attack. The pollock's reproductive prowess and fast growth make it a logical choice for an industrial fishery.

These days, Alaska's pollock support a fishery worth more than a billion dollars, but it wasn't always so. Just a generation ago, it was virtually unknown as a food fish. There's a story, recounted in fishing journalist Brad Matsen's book *Fishing Up North*, that may be too salty to be true, but we can't resist repeating it. According to one of Alaska's first pollock trawler captains, Muhammad Ali once impetuously bought 50,000 tons of frozen pollock from a broker in Caracas, convinced that it could be cheap protein for poor neighborhoods in New York. But no one would buy the unknown fish. It languished in a storage freezer, forgotten.

Today, people eat pollock by the millions of tons whether they know it or not. Unlike salmon, which is proudly presented with the names of the country and even the river where it came from, pollock is usually served under a different name altogether. You may recognize one: McDonald's Filet-O-Fish. Other times, pollock is sold as an entirely different species. It's usually the main ingredient in surimi, the Japanese fish paste that can be molded into any number of other ocean critters, like crabs, lobsters, and shrimp. The mild flavor of the imitation shellfish means the possibilities are endless. Pollock is often the anonymous white fish in the breaded fish sticks in your kid's school lunch, too.

In the 1960s, the Japanese ran the fledgling pollock industry in the Bering Sea. A couple of decades into Alaska's statehood, however, residents became annoyed that international interests were profiting from their fish. The marine exclusive economic zone, or EEZ, was established in 1976 with the Magnuson-Stevens Act, the United States' first comprehensive ocean stewardship policy. (The law was named for longtime Alaska senator Ted Stevens and Warren Magnuson, a senator representing Washington State.) The act allowed American companies alone to exploit oceanic resources within 200 nautical miles of the coast. While the act had nationalistic undertones, in truth, our ocean economy is still quite international. Fish caught in Alaskan waters can still be shipped directly to South Korea, China, or Vietnam for packaging rather than being processed in American-staffed onshore facilities.

Still, the Magnuson-Stevens Act spawned the modern American pollock industry. It started modestly enough with a single trawler, the *Storm Petrel* (the captain of which was the source of the possibly apocryphal Muhammad Ali story), steaming out from Kodiak in 1981. It exploded into an industry that in 2011 caught 1.15 million tons of fish worth $282 million, which quadrupled into more than $1 billion once the fish were processed and sold as surimi, Filet-O-Fishes, and fish sticks.

The modern pollock industry, while by no means perfect, is one of the best managed in the world. The total catch for the year is usually accomplished in a matter of weeks, reducing its impact on the marine environment. Most pollock trawlers have a government observer on board to count and report bycatch as it happens. This level of attention and information flow means that when the fish population dips—as it did in 2009 and 2010—fishery managers can move in to adjust catch rates and prevent any further collapse. After just a couple of years of catches of less than a million tons of fish, the pollock catch surged again in 2011. Data-driven, science-based management: It shouldn't be such a novel idea. Again, this isn't to say that pollock management has a spotless record. Besides chinook, the fishery has long threatened Steller sea lions by removing too much of their food. One year, a federal judge even shut down fishing to protect the sea lions.

The chinook salmon bycatch in the pollock fishery, however, is a notable black eye. The pollock fleet kills salmon that could feed one of Myron Naneng's

Yup'ik constituents or be sold in Anchorage or Tokyo or Miami for $17 a pound. And this "wanton waste" program, as Naneng called it, is perfectly legal.

OF THE THREE TENETS of responsible fishery management—set science-based quotas, protect habitat, and reduce bycatch—the pollock fishery did two of these things fairly well. After watching the salmon bycatch climb, Oceana's staff in Juneau began working to set a hard cap on the number of chinook killed by the fleet. At the time, the pollock fleet was catching more and more each year, reaching an all-time high of 130,000 chinook in 2007. That's nearly as much as the entire Alaskan Native community's catch.

After some good campaign work by our staff and other concerned groups, including a public meeting in front of the North Pacific Fishery Management Council in 2009, where dozens of Native and commercial fishermen, conservationists, and restaurateurs spoke passionately about protecting chinook, we got some of what we wanted. That year, the council set a hard cap on chinook bycatch at 60,000 salmon per year in the Bering Sea. A hard cap means that the pollock fishery will be shut down entirely if it surpasses that number. And in 2011, a hard cap of 25,000 salmon was set for the Gulf of Alaska to the south.

Those numbers may still sound high, but the motivation the hard caps provided has been huge. Since that disastrous year of 2007, the Bering Sea pollock fleet's bycatch number has dropped like a stone. In 2012, just 8,000 chinooks were caught by the fleet, the lowest number in more than a decade.

Native Alaskans have been eating salmon for thousands of years, using many of the same techniques to catch and prepare the fish as they do today. If we're going to feed the world, subsistence fishing like that still practiced by the Alaskan natives can't be steamrolled by powerful commercial enterprises. Because there is more than one definition of "subsistence," it's impossible to know exactly how many people are living directly off the water around the world today. The same is true of artisanal fishing, which encompasses a variety of traditional fishing techniques for commercial sale, such as the Senegalese tuna fishery, which exports yellowfin and skipjack caught from flat-bottom pirogues and employs 600,000 citizens in the West African nation. Artisanal and subsistence fishers around the world catch 30 million tons of fish for human consumption, the same amount as the industrial fisheries. But artisanal fisheries employ approximately 12 million people, about 25 times the number working in the industrial fisheries, and use an eighth of the fuel. Feeding the world from the ocean's bounty has to account for these fishing families, who individually cannot compete with big international fishing ships. When states and nations make decisions about fishing management, they must remember people like Myron Naneng. These people, too, have a right to the fish.

CHAPTER 9

A Philippine Story

"Be Aware"

Come on friends,

Let us all unite,

Let us save our seas,

That need to be developed,

Not just for today

But for your future my brothers and sisters

Stop dynamite fishing, my friend,

Put an end to using poison in fishing,

Do not throw your wastes on the shoreline,

So that mangroves, corals, and seagrasses may live.

Brothers and sisters, let us think of caring and nurturing
them for the future generation!

—BENJAMIN DELLOSA

A HEMISPHERE AWAY from Alaska, another of the world's biggest fishing
nations also faces enormous challenges in living up to the promise of good
ocean management. The Philippines is a nation of islands, more than 7,000 of

them, scattered over 2.3 million square miles of incredibly biodiverse ocean. This archipelago makes up the northern tip of the famed Coral Triangle. Coral reefs, along with mangrove forests and seagrass beds, provide some of the world's best habitat for fish and other marine life.

It's no surprise that the Philippines' culture and history are deeply tied to the sea. "A Filipino without fish is incomplete," as one resident put it. The country is the world's 11th-largest catcher of fish and 9th-largest consumer of it, eating some 2 million metric tons a year. Seafood composes more than half of the animal protein eaten by Filipinos, and fishing provides direct income to 1.3 million small fishers and their families.

The country has a greater number of marine protected areas (MPAs) than any other in the world: 1,600, most of them small and established in the last 2 decades. The country's governance is highly decentralized, and local jurisdictions are responsible for overseeing any protected areas within 9 miles of shore. While this can mean that citizens are motivated to protect their local no-take zones, only one in five MPAs is actually meeting its goals, which are twofold: to protect reef habitats from destructive fishing practices and to help reduce overfishing.

These are two of our three tenets of good marine stewardship. But the Philippines, one of the world's fastest-growing countries, with three times as many citizens as in 1970, has been struggling to maintain healthy fish populations in its seas. A third of the commercial seafood species in the country are suffering from overfishing. How some towns have taken up the mantle of responsible fishery management is our next story.

A POEM HANDWRITTEN in neat capital letters on poster board hangs on the wall of the guardhouse in Lanuza Bay, an inlet on the northeastern shore of Mindanao in the Philippines. It's low tide, and the seagrass meadow that blankets the ocean floor here undulates hypnotically in the clear water.

The guardhouse, a plain wooden structure raised 10 feet above the bay's shallow waters, sits just inside the edge of the Burgos Birds and Fish Sanctuary, a 165-acre MPA marked off with concrete pylons peeking above the bay's waters. You can reach it in an outrigger canoe with a motor in 20 minutes, but many of the people who live in the nearby town come in dugout canoes pushed along with bamboo poles. Simple as they are, the canoes represent jobs and food to the people who live here on Mindanao, the Philippines' second-largest island. The vast majority of families in the coastal towns here depend upon fishing for their daily meals and sell what remains at market.

Inside the guardhouse, Benjamin Dellosa helps translate the poem on the wall for a pair of foreign visitors. He's not just the author of the verse. He's also the founder of the fish sanctuary. In 1997, he realized that the fishers, who had once come in from the bay with 40 pounds or more of rabbitfish, groupers, and jacks, were now coming back with just 1 or 2 pounds and sometimes nothing at all. Anything the fishers caught went to feed their families rather than to market. For people living from day to day on the sea's bounty, the rapid decline was devastating.

Dellosa, a middle-aged man with a modest demeanor, isn't a fisherman. He's a rice farmer. Still, in 1998, he began patrolling the waters near Burgos overnight by himself. It was a dark, dangerous, and lonely assignment. Fishers often used makeshift bombs made from soda bottles and fertilizer to stun fish so they floated to the surface. This devastating technique may have offered up a quick bounty, but it destroyed the coral reefs and seagrass beds that fostered young fish. As a result, fishermen caught fewer and smaller fish in the local waters. Dynamite fishing was outlawed in the 1970s, but a lack of enforcement meant that it still happened frequently. Dellosa knew he was taking a risk by becoming a one-man enforcement team. "I put my trust in God," he said.

After 2 months on his own, Dellosa took some wine to his friends and convinced them to help patrol the municipal waters. An officially designated no-take zone followed. Inside the guardhouse, there is no furniture or kitchen— just a single plain room surrounded by a narrow covered balcony. There is a logbook with the names of all the locals who have taken shifts manning the guardhouse, some posters describing the fish species living in the waters, and Dellosa's poem. Now president of the MPA association for the bay, Dellosa provides a bag of rice to the volunteers at the guardhouse to ensure that the waters will not go unwatched again.

"I prefer to do some good for the people," he says in careful English. "I might die, so people will remember me."

The towns of Lanuza Bay are not large or rich. The smaller ones may have a few dozen households of raised thatch-and-bamboo huts peeking out from the riot of coconut and palm trees. Even the bigger towns, with their grassy central plazas and handful of municipal buildings, are home to about 30,000 people at most. The fishermen work out of canoes alone or in pairs, throwing nets or dropping hooks and lines. In the shallowest waters, sometimes they use corrals made of bamboo and mangrove stakes to entrap wandering fish schools. Compared with the massive fleet of industrial trawlers clearing out the high seas, it's hard to imagine this hyperlocal effort wiping out the fish in the rich seagrass beds, mangroves, and coral reefs that line the inner bay.

And yet the fish are, without a doubt, greatly reduced or gone from the coastal waters. In numerous Lanuza Bay *barangays,* the name for the village-level government, fishermen tell the same story repeated by Benjamin Dellosa: plenty of parrot fish and rabbitfish 30 years ago followed by a drastic decline. Lanuza Bay and its fishermen could stand in for many of the Philippines' coastal fishing communities. Many of the Philippines' poorer people are fishers; four out of five fishing families in the country live below the poverty threshold.

Imagine it: a small seaside village composed almost entirely of fishing families. Few residents are educated beyond a grade-school level. Almost everyone is devoutly Catholic, and nearly every woman has five or six children. The village is built on red clay that can't grow much besides coconut palms, so any food that doesn't come from the sea is bought at a market a half-hour motorbike ride away on a treacherous mud-and-gravel road. The ocean is the main provider of sustenance for these families. And when the men go out in their canoes and come back without fish, or the women can't glean crabs and mussels from the seabed at low tide, the family suffers. That means that there's huge pressure to fish illegally inside the sanctuaries.

So how can we reverse this situation? Not every town has a Benjamin Dellosa, willing to leave his wife and children to watch over the sanctuary alone.

St. Lucia, a thumbprint of an island in the Caribbean Sea, is a world apart from the Philippines. But a methodology that could help save the Philippines' coastal fisheries started here—with a bird, not a fish.

The St. Lucia parrot is a vibrant green bird with a cobalt blue face and ombré breast in shades of pink, purple, and yellow. Well over 1 foot long from beak to tail, the parrot flits among the branches of the island rain forest's upper canopy. By the 1970s, however, the Jacquot, as it's known locally, was in dire straits. Parrots were shot and sold into the pet trade or eaten, and the ones that evaded capture struggled to survive as the rain forest was cleared.

In 1977, a young British biologist named Paul Butler came to St. Lucia to study the 100 or so parrots that still existed. Most scientists had already written off the species to extinction. Butler and his colleagues spent a season observing the birds, wrote up a list of recommendations for the St. Lucia forestry department, and left.

Thankfully for the Jacquot, a wise person at the forestry department decided it wasn't time to give up on the parrot just yet. Butler was invited back to implement the recommendations he had made. They included some commonsense ideas: Increase the fine for engaging in the illegal pet trade, set aside some habitat for the birds to breed, and make the birds profitable by setting up tourist hikes to see them. But Butler's real innovation didn't come from a set of new rules for parrot conservation. The materials he initially wrote up, with a typical science-filled, reason-oriented rationale for saving this bird, made sense from a logical point of view. But they didn't get to the heart of why poor St. Lucians, who in the short term needed the money a bird smuggled into the pet trade could provide, should save the parrot.

"The people of St. Lucia had to think they had something special. They thought it was just a parrot, the same parrot as anywhere else," Butler said. "So the focus was to go to the villages and show that this was a unique species. What resonated was this sense of something they could be proud of, something that they had that no one else did. America could land on the Moon, but it didn't have the St. Lucia parrot."

Butler's Promoting Protection through Pride campaign worked. The parrot was named the country's national bird in 1979, the same year that St. Lucia won independence from the United Kingdom. Today there are more than 500 St. Lucia parrots. You can pay to go on the rain forest walk that Butler initiated more than 30 years ago and see the parrot for yourself.

The pride approach was adopted by Rare, a US-based group where Butler is now the senior vice president for global programs. It has been used to help save everything from finless porpoises in China to Siberian tigers in Russia.

Now, Butler and Rare are taking the pride campaign to coastal fisheries like the ones in Lanuza Bay. This is a big departure for the organization because the prior campaigns, with their emphasis on endangered and charismatic wildlife, hewed closely to the classic environmentalist focus on preserving biodiversity. But the rabbitfish (a species that feeds on fleshy algae, hence its name) that dart among the seagrass thickets of the Burgos fish sanctuary don't inspire awe like a tiger. And you don't protect a tiger just so you can later feed her cubs to your children (we hope). Rare's new fishery campaigns ultimately aren't about the rabbitfish at all—they're about the families that absolutely depend upon those little reef fish for tomorrow's breakfast.

After working on biodiversity-focused campaigns in the Caribbean, China, and Latin America, Rare began its fisheries campaign in 2010 with a cohort of conservation fellows in the Philippines. When chosen, the fellows were already working with a local agency. Rare sponsors each fellow in a two-year campaign that essentially trains him or her in marketing and media. Each fellow earns a master's degree in communications from the University of Texas–El Paso, and the community gets a custom-designed campaign to protect its ecological heritage. Rare's hope, of course, is that the programs will continue even after the 2-year campaign is completed, and the organization provides funding for alumni who continue their work.

In Lanuza Bay, there are three Rare conservation fellows. Each one works with one to three *barangays* to improve enforcement in the MPA with a marketing campaign that includes a slew of T-shirts, banners, and billboards, capped off with a colorful campaign mascot designed with input from the community. Each mascot is an anthropomorphized signature marine species of the barangay: an oversize rabbitfish named Rabita, a giant clam named Tilan, a blue lobster named Lob-Lob. The fuzzy mascots might seem cloying to Americans used to being bombarded by marketing pleas. But watching them in action, it's impossible to deny the mascots' effectiveness: They are swarmed by children whenever they make an appearance, and adults line up to have their photos taken, too. The mascots show up in mayors' offices and get invited to all the towns' events. In these small villages at the edge of the Pacific, Rabita the rabbitfish might as well be Mickey Mouse.

The point of all this—the mascots, the T-shirts, the billboards—isn't just to instill local pride, of course. It's a clever way to get people to understand and support healthy fisheries without having to endure the doomsday talk of environmentalists or the academic detachment of fishery managers. It's a good match for the Philippines because of the country's decentralized government, which leaves the local towns to protect their own waters.

VINCENT DUEÑAS, the Rare conservation fellow for two barangays in Lanuza Bay, is leading a brainstorming session at the guardhouse on the rocky edge of the Uba Fish Sanctuary. A charismatic lay minister with a disarming smile, Dueñas was already working in coastal fish management in the mayor's office before teaming up with Rare in 2011.

A dozen fishermen and their wives sit on benches on the guardhouse's balcony and watch attentively as Dueñas shows a printed draft of a billboard. It displays a hand-drawn image of a guardhouse against a backdrop of coconut trees. In Visayan, a local language that also includes the occasional English term like *billboard* or *illegal*, Dueñas asks the fishers what they're looking at and, with the subtle hand of a minister, guides them through a conversation about the depiction of the MPA and its meaning to the community. Eventually, the group, which was once quiet and observant, is fully engaged. The fishers point animatedly as they speak rapidly in Visayan. They suggest moving the guardhouse in the image closer to shore, as the Uba guardhouse is located on a rocky promontory. They want a man in the patrol boat in the water and the MPA's name on the boat.

Paul Butler observes from one corner of the balcony. He has traveled from his home in the south of England to check in on Rare's three Lanuza Bay projects. As the fishermen discuss what the billboard's message should be, Butler—who's naturally exuberant and easily commands an audience, especially when he's the only foreigner for miles—keeps quiet. "Regardless of the wording, it's theirs," he says. "This is probably the first time someone's asked their opinion."

Indeed, the fishermen are rapt. It has been only a few months since Dueñas started his pride campaign, and the guardhouse logbook shows that the no-take zone has had round-the-clock volunteers since he began his efforts. A team of women watches during the day, while men take the long shift overnight. Since they started 24/7 guarding in earnest, there hasn't been a single illegal intrusion into the MPA.

One of the volunteers is Eleuterio Sanchez, a 65-year-old grandfather of 27 with a deeply lined, kindly face. A lifelong resident of Uba, Sanchez has been fishing since childhood like his father did before him and his sons do now. He wore his new black "Fishery Enforcement Team" T-shirt to Dueñas's meeting. He remembers when dynamite fishing and killing fish with shrimp bait laced with poison were common practices. These damaging techniques resulted in lean years when there weren't enough fish to make a living, and he turned to growing rice or cassava, a root crop, for food. But he always returned to the sea. Since the Lanuza Bay no-take zones were established in the 1980s and '90s, Sanchez has seen the return from the sea improve. Now, he catches about 15 pounds of fish a day.

And he sees the value in Dueñas's work, too. "Even at this age, I still guard so the children will see this is the right thing to do so they will not be illegally fishing, too," he says through a translator.

⟨fish ornament⟩

SO IS RARE'S town-by-town approach a success? Protecting a handful of small villages with postage-stamp-size no-take zones doesn't seem on its face like enough to save the world's oceans and its seafood. The Philippines has the most MPAs in the world by number, but only one in five is well managed. The country will have to improve its track record because its population of seafood lovers is skyrocketing; there are three times as many Filipinos today as in 1970. And the local fishermen in outrigger canoes represent the Philippines' poorest population. They are the Filipinos who are least likely to have access to indoor plumbing and clean water and the most likely to have large families with half a dozen kids.

Filipino fisher families like those in Lanuza Bay will continue to struggle for one big reason: They alone are not responsible for the declining amounts of fish to be caught in Philippine waters. No, the local fishers in dugout canoes must compete with the well-financed, technologically advanced industrial fishing fleet. These industrial boats are part of the same industry that has wiped out fisheries around the world, as described in Chapter 3. And most devastating of all is that they are catching the same fish the local fishers are trying to catch to feed their families. Of the top 10 species the local fishers and the commercial industry each catch in the Philippines, 8 are the same. In the 1950s, local fishers caught 70 percent of the Philippines' total fish catch. Today, it's just 46 percent—even though commercial fishing employs a fraction of the number of local fishers.

And then the numbers don't tell the whole story. In the Philippines, artisanal fishers are defined as boats that weigh less than 3 tons. These are the only boats allowed into "municipal" waters, which are the shallow, reef-filled bays like Lanuza. The dugout and outrigger canoes used by the locals in Lanuza Bay fall well within this category, but some commercial operators have found a way to skirt the rules. They've developed a large fleet of *mini-trawlers* that are under the 3-ton limit, but they don't use them like the local fishers. These mini-trawlers are not fully accounted for in the country's government statistics, and they mean that the true artisanal fishers are even more threatened than the numbers indicate.

Officially, the number of industrial fishing vessels in the Philippines tripled between 1980 and 2002, but this intensified race to catch more fish only temporarily results in increased catches. Eventually the fish populations fold under the pressure. According to the FAO, the Philippines' total annual fish catch leveled off more than 20 years ago at about 1.82 million tons. And some of those catch statistics are a total fantasy. During Ferdinand Marcos's 2 decades as president, the Philippines reported jumps in catch every year, despite there being fewer and fewer fish actually in the water.

Given that the country has some of the world's richest seas, this is an astonishing indictment of the "faster, farther, deeper" attitude of the commercial fishing industry. Catching fewer fish each year isn't exactly economically sound. Scientists have estimated that overfishing costs the Philippines $125 million a year.

The increasing scarcity of the Philippines' fish means that seafood, which was once a working-class meal, is now increasingly priced out of its grasp. In Cebu, the oldest city in the Philippines and an island-hopping plane ride from Mindanao, the customers at the downtown fish market are primarily tourists from Manila because the fish are too expensive for locals to buy. Rabbitfish, the mascot species for Vincent Dueñas's campaign, costs $5 a pound in Cebu. That's a whole day's gross earnings for a fisher in Lanuza Bay. As their backyard fish become scarcer, many of the coastal families, which have few other economic opportunities, will risk going hungry. Fifteen percent of the Philippines' population is already undernourished. It's among the top 10 low-income, food-deficit countries. These countries produce a third of the world's fish but have three-quarters of the world's undernourished kids.

The Philippines will have to establish strong national and regional policies on science-based quotas, habitat conservation, and less bycatch to better manage its out-of-control commercial fishing industry and protect its citizens in places like Lanuza Bay. But given their limited resources, the people of these villages—with support from Rare—are improving their chances. No-take zones, even small ones, are proven to work. Scientists have confirmed that MPAs have greater overall biomass, higher biodiversity, and bigger fish than outside the no-take zones.

Because the Philippines is located in one of the world's most diverse and productive ocean ecosystems, the health of its waters is critical. Right now, the country's MPA program is fragile. The federal government must greatly reduce overfishing of the Philippines' 2 million–plus square miles of national waters by the industrial fleet in order to protect the coastal citizens who rely on the ocean for sustenance. Right now, the barangays are shouldering most of the responsibility, and they have seen some town-by-town success. Yet they can't do it on their own forever, or the commercial fleet will undermine all their efforts.

Good coastal fisheries stewardship has been shown to alleviate poverty. And local success should beget national success: The creation of local MPAs helps provide the grassroots muscle for defeating the national policies that favor the industrial fleet and drive ocean depletion. We could see a country with a healthier ocean that supports the people who depend on it. That's a beautiful thing.

CHAPTER 10

How to Save the Oceans and Feed the World

Never before has man had such capacity to control
his own environment, to end thirst and hunger.

—JOHN F. KENNEDY, 1963 ADDRESS BEFORE THE UNITED NATIONS

IN 2009, WE interviewed marine scientist J. E. N. Veron, or Charlie, as he's
called by everyone who knows him, for Oceana's magazine. An Australian who
has spent more than 7,000 hours underwater studying coral reefs during his
50-year career, Veron turned philosophical when we asked him what was the
best thing he'd ever seen under the waves.

"So many amazing things. Great big monstrous things, it would be
impossible to say," he said.

*Aboriginal fish traps found ten meters under the surface. Caves where
aborigines once lived. Really big tiger sharks or whale sharks, whales.
Being sounded out by humpback whales is the most terrifying
thing—the blast of sound they use to work out who the diver is is quite
incredible. I've been diving so much and seen so many things that
it's all off the scale. It's very sad that young people today will not see,*

and have no chance of seeing, what I've seen. It makes me enormously sad. I have two young children; they will never see what I've seen. No one will.

Charlie is a leading scientist and genuine ocean conservationist. His story illustrates one familiar way the ocean engages some of its most passionate defenders. It's clear that if a substantial share of the world's influential people were avid scuba divers, ocean wildlife would already be protected and its long-term abundance assured. Or would it?

Land-based conservation has often made progress by focusing people's attention on special creatures. By featuring bears, tigers, elephants, and other charismatic megafauna, advocates are able to harness the popular clout they need to successfully push policy makers to take action. Of course, the challenge for a campaign to broadly protect biodiversity is to find a way to make sure popular affection for these poster animals results in real protections for the vast numbers of creatures without the big eyes that so strongly invoke our instinctive emotional response.

Obviously, there are big ethical problems with biodiversity policy making that's driven by the limpidness of creatures' eyes. A world in which we save only the cute is not worthy of us. But a practical person faced with the necessity of stopping ocean collapse can reasonably insist that we learn what lessons we can from the marketing of land conservation.

Happily, the ocean does indeed give us charismatic creatures: marine mammals. We can't resist whales, dolphins, otters, and polar bears. We identify with them even if we will never don a scuba tank and learn to breathe through a regulator. They are mammals like us. They breathe air, give birth to live young, and nurse them with milk. They are social. In their own ways, they talk and sing.

They are everything fish are not.

When we started Oceana a little more than 10 years ago, we had a meeting at which we discussed replacing our dolphin-inspired logo with one based on a flounder. Some of us wanted the anti-charisma of the flat, floppy, bottom-hiding food fish to be the image on our calling card. Other voices chorused against the change. We stuck with the dolphin.

It's tough to even come up with an example of a popular conservation campaign celebrating the life of a terrestrial animal that we eat. The pork producers, chicken industry, and cattle raisers promote their meat, not the animals. Their advertising shows us deliciously prepared food on a plate and conspicuously avoids reminding us of its previous identity as a living creature. In marine conservation, there is no equivalent to the photo of the beautiful amber field of wheat or the green orchard that adorns the cereal box or orange juice carton. So engaging people in saving the ocean's creatures because we need to eat them takes us into brand-new territory.

It can seem, at first, like an impossible and contradictory challenge. But drop your land-based expectations, and you'll see that we can do it. Have you ever cooked a live lobster? Have you ever taken a wooden mallet to a crab? Have you ever gone fishing and caught your own dinner? Have you ever opened a live oyster and eaten it? Have you ever gone to a sushi restaurant and had the chef cut the fish fillet in front of you? Or what about a Chinese restaurant where you chose the fish for your meal from an aquarium?

We already associate our love of good seafood with the living ocean creature. It's a long-standing, familiar part of the culture of seafood restaurants, where fish can be ordered whole, skin on, and even of our grocery stores, where the fish counter presents us with fish whole on ice, recognizable as animals once swimming in the ocean. We don't flinch. We are reassured by this tangible connection to a natural, healthy, fresh, wild food animal.

Building on that foundation, we can create a new relationship between ourselves and the creatures of the sea. Till now, our relationship with the animals that feed us has had only two names: We are their farmers or we are their hunters. If we are going to rely on them forever as a source of healthy protein for billions of people, the fish in the sea call us to a different relationship—the one called stewardship.

Stewardship demands understanding that something is fully entrusted to one's care. It's not the same as ownership. A steward is a manager who is highly ethical and cares what happens in the long run. A multigenerational orientation is built into the term from its origin in the Middle Ages, when feudal lords appointed estate managers they called stewards. Stewards are operating productive enterprises, not protecting pristine nature zones. But they are doing so with an awareness of their responsibility to perpetuity. They have control, they are actively engaged in making key decisions, and they are accountable for outcomes. In short, they are ethical managers.

So can we be stewards of our oceans? Do we have the practical conditions we need—the authority analogous to that delegated by the feudal lord—to actually do this for our oceans? If the oceans are legal high seas in which no one is in charge, how can we possibly hope to implement ocean stewardship?

In the 1980s, coastal nations began establishing 200-nautical-mile exclusive economic zones off their coasts. This means that all fishing in that zone is conducted only under the management rules of the closest coastal nation. These nations set the rules for fishing in their zones just as they do for fishing in their rivers and lakes. Since strong and effective action by the world's international bodies is quite rare, the formalization of national control of ocean zones is encouraging. It creates the opportunity to save the ocean country by country.

This change has not been easily absorbed because it's revolutionary. Before the 1980s, most nations governed just the ocean waters 3 to 12 miles off their coasts—a distance, it is speculated, founded on the length of a cannon shot. The ocean beyond that splashing iron sphere was considered international property. That meant a fishing free-for-all, a phenomenon known more eloquently as "the tragedy of the commons." This concept was first laid out by the economist Garrett Hardin in the late 1960s. It refers to the once-familiar practice in rural England of grazing all the village's livestock together in a common space. It sounds like an idyllic arrangement, but the paradox pointed out by Hardin was that no individual villager was motivated to protect the common land from overgrazing and ultimately destruction.

In the oceans, the phenomenon explains overfishing in the high seas beyond the exclusive coastal zones as a predictable consequence of international fishing fleets pursuing their self-interest. The commons of the high seas allows fishing fleets to fish as hard as possible. That's because each country gets the immediate benefit of *all* the fish its fleet captures, but *shares* the long-term benefits that arise from all the countries operating under sensible catch restraints. This means the rewards for aggressive short-term exploitation are much larger than those for sensible management. In the absence of enforceable international coordination, every nation sees it the same way and competes for the same resources. This results in fishery crashes that hurt everyone. Newfoundland cod, Mediterranean tuna, Chilean sea bass: These are all victims of this story. It benefits no fishing fleet to destroy its own livelihood, yet the historic record is replete with such disasters.

But we have reason to be hopeful about the future of our wild seafood supply. Those 200 nautical miles of national waters are the key. That's because most ocean life dwells in those coastal areas. These regions contain 99.9 percent of the planet's coral reefs. They also provide 7 out of every 8 pounds of wild fish caught in the ocean. The high seas, by comparison, are more like desert mesas: difficult and dangerous to reach, and the catch often is not worth the cost of the fuel to get there. High-seas fishing is financially feasible only because of the $6 billion in annual fuel subsidies the fishing industry worldwide receives from governments each year, meaning that in many countries, you're paying for your fish in two ways: first as a taxpayer and second as a consumer. (In the United States, the government has mostly moved away from direct subsidies that promote fishing capacity, like those for fuel and other operating or capital costs. However, the United States does fund fishery research, management, and regulation with general taxpayer funds rather than taxes earmarked for the commercial fishing fleet.)

The way marine life clings to the coasts is good news for conservation because it means we can protect the oceans nation by nation rather than depending on international bodies like the United Nations or the many bureaucratic entities that oversee the big predator fish in transoceanic fisheries. (The International Commission for the Conservation of Atlantic Tunas, or ICCAT, to name one, is such a model of inefficiency and ineptitude that marine scientist and conservationist Carl Safina calls it the International Conspiracy to Catch All

Tuna. The name has stuck, unfortunately so for bluefin tuna, among the most threatened seafood species in the world.)

The three steps for seafood sustainability that we've outlined—set science-based fishing quotas, reduce bycatch, and protect habitats—are perfectly achievable goals for the nations that control the world's biggest fisheries. And each nation can interpret these tenets in a way that works for the ecological and cultural uniqueness of the country. Some nations, like New Zealand, adopted individual fishing quotas that can be bought and sold like stocks by fishermen and fishing companies. Chile portions out catch shares over specific patches of seafloor where artisanal fishermen gather mollusks. The specific methods used to set quotas, protect habitat, and reduce bycatch vary, but everywhere they are sensibly enacted and enforced, they achieve the same improvements in the water.

And that's one of the greatest things about ocean conservation: The ocean is astonishingly fertile. Ocean fish are very resilient creatures. Some of them lay eggs by the millions. You enact and enforce smart fishery policies, and the fish come back. You can see results yourself, in your lifetime. It's simple and raw and beautiful.

In southwest India, for example, several thousand fishermen still practice the ancient method of hand-gathering short-neck clams in the Arabian Sea. When the fishers and the district government saw that the clams were dwindling, they ended harvesting during spawning and spat season and established a minimum net mesh size to ensure that the youngest clams could escape. Between 1996 and 2002, the clam haul increased sevenfold.

And in the Sea of Japan near Kyoto, the snow crab population crashed in the 1970s after peaking at 550 tons. Juvenile snow crab bycatch in the off-season amounted to up to 60 percent of the population. Things started to turn around when managers set aside protected areas for spawning in the 1980s, even placing large concrete blocks on the ocean floor to ensure that no one would try to trawl in the closed areas. Catch limits were finally set in the 1990s, and the fishery as a whole moved away from trawlers and toward Danish seine nets that don't damage the seafloor quite as much. The snow crab catch has stabilized at about 110 tons after dipping to 86 tons in the 1970s.

In New Zealand, widely considered one of the world's most sustainable fishing nations, the country's largest fishery has been pulled back from the brink of collapse. The catch of hoki, a whitefish in the hake family that's often sold at fast-food places like Long John Silver's, declined from 275,000 tons a year to 100,000 tons in 2007. It's since been gradually recovering. New Zealand, which has one of the world's largest ocean territories, adopted a revised national quota management system. The country's fleet was reduced by more than half, and the value of the fishery (i.e., the market value of the fish caught each year) has tripled to more than $3 billion. As one fishing company manager put it, "Fish are getting easier to catch, and we catch them quicker."

In Alaska, the snow crab fishery was run derby-style 20 years ago: Fishermen snatched up as many crabs as possible, often hitting the quota for the season within a few days. The race made the work more dangerous than coal mining, with crews working around the clock in freezing conditions to beat their competitors to the crabs and often overloading boats to the point of capsizing. By the turn of the millennium, catches fell from 92,000 tons in 1999 to 13,000 tons in just 1 year. By 2012, after a decade-long rebuilding plan that included greatly reduced catch quotas, the snow crab had largely recovered. Today, a catch share program has reduced the number of vessels and fishermen, and the fishery has little resemblance to the derby days of past.

In New England, a lesser-known cousin of the cod is a whitefish called haddock. Just like cod, it was heavily fished. In 1930, 37 million haddock were caught while another 70 million to 90 million baby haddock were discarded at sea as bycatch. The fishing continued at this breakneck pace until the 1970s, when trawlers from East Germany, Poland, Spain, Japan, and other countries "discovered" haddock and helped to push the fishery over the edge into collapse. In response, the federal government enacted regulations that closed certain areas of the ocean that were important haddock habitats and limited the number of days fishermen could target haddock and other groundfish such as cod.

While cod has struggled to return, with a population that's still just 10 percent of a healthy number, haddock has come back in spades on Georges Bank, where haddock "baby booms" have spurred recovery. Now the New England Aquarium calls haddock an "ocean friendly seafood choice" and the Monterey Bay Aquarium Seafood Watch considers haddock to be a "best choice" or "good alternative."

In Europe's North Sea, the groundfish fishery peaked after centuries of exploitation with the advent of the bottom trawl in the late Victorian era. Within a decade, however, England's fishing ports were bringing in fewer fish every year despite the efficient technology of the trawls. North Sea groundfish once fed some of Europe's oldest cities; these days, only about 16,500 tons of cod come out of the sea each year. But the North Sea cod may be on its way back. Strict quotas, seasonal closures, and fleet reduction by way of decommissioning many of the boats have allowed the North Sea's cod biomass to double in the last 6 years. In the Baltic Sea, the North Sea's brackish eastern neighbor, the spawning biomass of cod has grown sixfold since the 1990s after authorities cracked down on illegal fishing.

On England's southern coast, fishermen in the town of Brixham are trying hard not to repeat the mistakes of the country's northern fishing towns. Brixham trawlers now use improved gear that's much more selective than its predecessors. The benefit is threefold: The marine habitat is less disturbed, less bycatch makes it into the nets, and fishermen make an extra $300 a week by fishing during the time they would have spent sorting their catch from the rest of the fish and rubble.

And farther south, off the coasts of France and Spain, fishermen are once more fishing for anchovy in the Bay of Biscay. Basques fished for sardines in the Bay of Biscay for centuries, but in the 1980s, they also began to target anchovies to keep the fishery alive. The delicious little anchovy was quickly overwhelmed. After groups including Oceana called for the enforcement of quotas that could rebuild the fishery, the European Union closed it in 2005. In 2010, scientists found that the fishery had rebounded to its 1987 level, and Basque fishermen are out on the sea fishing for anchovy once more.

Lastly, a tiny fishing town on the southwest coast of Madagascar in the Indian Ocean established closures to protect the populations of octopus that locals gathered from the seafloor during low tide. Thanks to the closures, Malagasy gleaners, many of whom are women, are catching twice as many octopuses as they were before. Neighboring communities saw the success and began their own closure programs, more than 50 of which have been implemented since 2004.

These stories have a common narrative: When sustainability measures work, they often begin with short-term pain. Overexploited fisheries have too many boats catching too few fish. Some fishermen end up leaving the industry, whether it's via decommissioning boats, reducing the number of individual quotas handed out, or old-fashioned bankruptcy. But once you get fishing pressure down to a reasonable level, the fish usually return. The fishermen who are left catch more fish, make more money, and create more jobs. It works on all scales, from the Filipino subsistence fishermen in Mindanao taking a few pounds of rabbitfish a day to the pollock fishery in Alaska, where millions of tons of fish come out of the North Pacific every year.

ANOTHER THING THESE stories have in common is that they show ocean conservation happening at the national level. There aren't any complex international negotiations going on here. Basic tenets of smart fishery management can have the same positive results on a big or small scale, but we want to save the *world's* oceans. You might deduce that we need all the world's countries to make this happen, so let's do a little math. There are 196 countries in the world; 44 of them are landlocked, so let's remove them. That leaves 152 nations with ocean-exclusive economic zones stretching 200 nautical miles from their coasts—still a big number. But if you drill down further and look at just the nations with the largest ocean territories, the situation clarifies. Just 25 nations control 76 percent of the world's coastal oceans. And just 10 nations control 51 percent.

That means more than half of the world's coastal regions are controlled by these nations: the United States, France, Australia, Russia, New Zealand, Indonesia, Canada, the United Kingdom, Japan, and Chile.

Let's look at it another way, by the weight of the world's wild seafood catch. This is a better measure of where the most productive fisheries are. The numbers tell the same story. The top 25 nations control 75 percent of the world's wild seafood by weight; the top 10, 53 percent. These top 10 nations are Peru, China, the United States, Russia, Indonesia, India, Chile, Japan, Norway, and Denmark. In order to save the oceans and provide the world with healthy, affordable, renewable protein, we don't need the cooperation of every nation. We need these key governments to implement the three basic tenets of ocean conservation. The top 10 countries look like a diverse group, and to be honest, not all of them have played well together in the past. Fortunately, for the most part, they don't have to cooperate to save the oceans.

And unlike a lot of other environmental issues, seafood isn't highly politicized. In the last decade, the world's largest marine reserves have sprung up under regimes from across the political spectrum: the 338,000-square-mile Chagos Marine Reserve in the Indian Ocean, established under the Labor government of British prime minister Gordon Brown in 2010; the 140,000-square-mile Papahanaumokuakea Marine National Monument encompassing the western Hawaiian Islands, approved under George W. Bush's administration in 2006; and the enormous 900,000-square-mile reserve created under Australian prime minister Julia Gillard, a leader of the Labor party, in 2012. In Chile, where Oceana has worked for the last decade, we've achieved many of our goals, including reforming salmon farming practices, establishing scientifically sound fishing quotas, and protecting essential fish habitat under conservative president Sebastián Piñera.

Oceana campaigns for these changes even in places where the government is not as supportive of ocean conservation as Chile's has been. We do this in a practical and focused way. We identify a policy change that will make a substantial contribution to improving ocean abundance—typically, scientific quotas, nursery habitat protection, or bycatch reduction. We determine which policy maker has the authority to implement that change. We then meet with

that policy maker and present the scientific data and legal arguments supporting the change. If the policy maker is unwilling to act, we design a campaign to
push him or her into action. This typically involves a mix of activities, including publishing scientific reports, mobilizing attention from the press, engaging
grassroots activists, employing in-person advocacy in the halls of government,
and, if necessary, taking legal action to compel enforcement of fishery conservation laws. To make sure that we are accountable for securing results, we give
ourselves 3 to 5 years to win.

Sometimes it's a battle. But usually it's not a narrowly partisan one with
all the Democrats on one side and the Republicans on the other, to offer a US
example. Fisheries conservation is bipartisan because, frankly, it just makes
sense. There's an economic argument for it: The World Bank has said that overexploited and poorly managed seafood stocks cost the world $50 billion a year.
There's a humanitarian angle: Fishing is a source of income and food for
200 million people, many of whom are in developing countries, like those
Malagasy octopus gleaners. Seafood is the most cost-effective animal protein in
the world, making it affordable for most people (as long as they're not eating at
Nobu). And saving our seafood has the obvious and ultimate conservation
angle: Without the billions of fish and tiny crustaceans that make up the middle
of the marine food web, we would have no dolphins, no narwhals, no orcas, no
monk seals, no albatrosses, no polar bears, no hammerhead sharks, no emperor
penguins. The fishing industry may always see these creatures as enemies of its
bottom line. So we'll have to remain vigilant to protect them, too, with a more
nuanced understanding than the last centuries of ocean exploitation had.

The 21st century will be the era when we push our planet's ability to
absorb human impacts to the limit. Honestly, even with nearly a billion people
hungry today, we've done a remarkable job simply producing enough food to
keep up with a global population that has doubled since the 1960s. The Green
Revolution allowed us to roughly keep pace with global demand for food. Our
major failing has not been in producing food. It's been producing it at too great
a cost to both the environment and the world's poor.

But the story may be different in the 21st century, when the population is growing at a rate more rapid than ever before. We are on target to add an additional billion people to the planet in 12 years. And the planet is under greater stress than ever thanks to climate change, which is pushing the world's breadbaskets toward drought. In 2012, the United States had the hottest July on record, eclipsing even the Dust Bowl days of the 1930s, while Europe experienced its coldest winter and wettest spring in decades. These extreme weather events are going to make it harder and harder for farmers to produce 50 percent more food, which is how much the United Nations says we're going to need by 2050. Global warming affects the oceans, too, because, as history has shown, they absorb more carbon dioxide than even forests do; they're the world's greatest carbon sinks. But the oceans are becoming so saturated with carbon that the pH level of the water is dropping, making it acidic. This threatens the corals and little microscopic sea creatures that use calcium carbonate to make shells. No one knows exactly how acidification will change the oceans or how fast. But combined with the warming waters, which are already sending fish migrating toward the poles, we know the oceans are going to look pretty different in another hundred years.

This means that we have to work extra hard today to ensure that the oceans are as healthy as possible as we head into a more challenging future. If we succeed—and it's entirely possible that we will—the marine bounty can be richer than it is now. According to a recent study in *Science,* if the world's fisheries were better managed, they could yield up to 40 percent more of the world's healthiest, most environmentally friendly protein: wild seafood. That would mean that 700 million people could enjoy a nutritious meal each day in 2050 because they would be eating the perfect protein.

AT CAPTAIN BOB'S, a breakfast and lunch joint behind the Blue Water Marina in North Bimini, fishermen start grabbing coffee and French toast at 6:30 in the morning before heading out for a day on the water. One of the regulars is Bouncer Smith, a charter boat captain with a wide belly and a wider smile. He helms a 33-foot Dusky for day trips on the Caribbean's turquoise waters. The 64-year-old Smith has been guiding fishing trips for everything from bonito to wahoo for more than 45 years. Based out of Miami Beach on most days, he's a South Florida institution.

One June morning, Smith walked into Captain Bob's. Someone asked him how he'd done during the full moon snapper spawning season, and Smith responded that his charter had caught four fish and gone home. "You must have caught 30," the man responded. "You're too good of a fisherman for that." Soon another patron wandered in and asked Smith the same question. Again he answered four fish. Finally, when a third patron expressed surprise at Smith's small haul, he got exasperated. "I said, 'Here's the fact of the matter. I don't fish the snapper spawn, generally,'" Smith recalled. "'But there's people who are overly greedy, and they see them coming in one after another and they want to catch all they can. I'm not going to be part of destroying the spawning population of those fish. I had a charter that was willing to keep one fish per person and went home. And we left fish there for the future.'"

Smith's speechifying shouldn't have surprised his fellow regulars. He has been a leader in conservation in the recreational fishing world for nearly as long as he's been on the water, winning commendations as the region's top tag-and-release boat for numerous years. He has worked closely with the Billfish Foundation to both tag and help conduct DNA testing on Florida's fish.

"And when I got done spouting off my opinion, almost everybody in the room was in complete agreement. And if there was anybody that didn't agree, they certainly weren't going to open their mouth, because they would have been severely outnumbered," Smith said with a laugh. "But it was very inspiring that so many people agreed that a fishery was overabused and people needed to respect the fish."

Respect the fish. It's a simple proposition. Leave enough in the water to renew the next generation and the oceans will reward you. The oceans are the source of all life, and they sustain us today, even though we know less about the deepest recesses of the oceans than we do about the surface of Mars. Those vast waters could provide even more sustenance for us in the future. But first we must meet our potential as good stewards of this blue planet.

DO YOUR PART

ONE OF THE KEYS to bringing the world's fisheries back from the brink is you. Here are some ways that you can help protect the oceans and wild seafood for the future.

- **Speak up.** The key to saving our oceans is better ocean policy. You can help make this happen by calling your elected officials and community leaders to discuss the ocean issues you care most about and writing letters to the editor of your local newspaper. And don't forget to vote—electing the right public officials is essential to good ocean policy.

- **Get smart about the issues.** Read about ocean issues. Stay informed about the latest news impacting ocean policy. *The Beacon,* Oceana's blog (www.oceana.org/blog), is a great place to start.

- **"Eat an anchovy"—consume sustainable seafood.** As we said: "Eat wild seafood. Not too much of the big fish. Mostly local." We have lots of great recipes for you to try, beginning on page 131. Stick with mussels, clams, and other shellfish; and try some sardines, anchovies, and other lower-on-the-food-chain fish (they're delicious, plus you'll be helping to create a more sustainable market for these important fish). Tilapia, catfish, and wild salmon can be great alternatives. You can find recipes, seafood guides, and more at http://oceana.org/livingblue.

- **Get smart about your seafood.** Talk to your local seafood purveyor, and get to know your local fishmonger if you can. Ask questions about the fish they sell. Request that they stock more sustainable options—like small fish—for you to cook with. And read our Easy Guide for Eating Seafood Responsibly on page 167.

- **Go to the sea.** Get your family and friends to the ocean. It's the best way to persuade someone to care about saving the seas. Fish responsibly, scuba dive, snorkel, sail, boat, kayak, paddleboard, or just take a walk on the beach.

- **Host a fund-raiser.** Host a sustainable-seafood barbecue fund-raiser. Grill up some mussels or sardines; watch some ocean movies like *Finding Nemo, Blue Planet,* or *Oceans;* and collect donations for your favorite ocean-related nonprofit.

- **Join Oceana.** Join the more than 550,000 members and e-activists in more than 150 countries who are already members of Oceana, the largest international organization focused 100 percent on ocean conservation. You can sign up to receive action alerts and more at www.oceana.org/join.

- **Spread the word.** Tell your friends and family what's going on with the world's oceans and what they can do to make a difference by joining the conversation on Facebook and Twitter.

EATING THE PERFECT PROTEIN

AS WE'VE DISCUSSED in these pages, appalling numbers of fish that could feed people are currently either discarded or ground up and fed to livestock. And it's even more of a shame because these fish—think anchovies, sardines, and mackerel—are actually quite delicious.

Prove it, you say? We asked some of our favorite top chefs to give us their best mouthwatering recipes for seafood that you typically might not eat. They graciously complied, and we are pleased to offer you the results in this section. We are also very grateful to our chef partners for their help.

These recipes are delicious, and all include fish that you can cook without guilt (mostly), following our rule of thumb for conscientious fish eaters: "Eat wild seafood. Not too much of the big fish. Mostly local."

We include lots of wonderful recipes for small fish—like Barton Seaver's sardines—and a few familiar favorites in the large-fish department, such as Cat Cora's halibut and Nora Pouillon's black cod. The more adventurous cooks will find some unusual offerings, like a jellyfish salad from Mario Batali and a dish featuring mullet from Dan Barber.

All of these recipes are sustainable, and they all showcase the perfect protein. Enjoy!

BARTON SEAVER'S SMOKED SARDINES
WITH HEIRLOOM TOMATOES AND HERBS

PREP TIME: 5 MINUTES | SERVES 4

Barton Seaver has been called the Alice Waters of seafood, and when you look at his cooking résumé, it's easy to see why. His status as a sustainable-seafood celebrity has even led to a foray into policy—he joined Oceana in urging Congress to pass legislation to curb seafood fraud. Chef Seaver notes, "In keeping with the ease of a tomato salad, all you have to do here is to open up the can of sardines and chop some herbs. Regular red tomatoes will also work, but heirlooms have interesting and unique flavors and give you the ability to mix and match different personalities on the plate."

2 pounds super-ripe heirloom tomatoes (preferably a colorful mix of varieties)
2 cans (6 ounces each) oil-packed smoked sardine fillets, oil reserved
Kosher or sea salt to taste
¼ cup chopped fresh soft-leaved herbs such as parsley and chervil
1 bunch scallions, thinly sliced, including the green tops

Core and slice the tomatoes ½" thick and distribute among 4 plates. Flake the sardines over the tomatoes and drizzle the oil over the top. Season with salt and sprinkle the herbs and scallions over all. Serve immediately.

ERIC RIPERT'S CLAMS WITH SPICY SAUSAGE

PREP TIME: 10 MINUTES | COOKING TIME: 20 MINUTES | SERVES 4

If Le Bernardin is New York's temple to seafood, then chef and co-owner Eric Ripert is one of the ocean's culinary deities. Chef Ripert learned to cook as a young boy in France, has since earned nearly every cooking accolade there is, and still has spare time for conservation—he recently joined Oceana in calling for increased seafood traceability in the United States. But don't let the chef's pedigree intimidate you; this recipe is a breeze.

¼ cup extra-virgin olive oil

3 cloves garlic, thinly sliced

1 onion, thinly sliced

1½ teaspoons curry powder

1 teaspoon finely grated lemon zest

¼ pound andouille sausage, thinly sliced

1 cup chicken stock

4 dozen littleneck clams, well scrubbed

¼ cup chopped cilantro

Lemon wedges for serving

Warm the oil in a large saucepan over medium heat. Add the garlic, onion, curry powder, and lemon zest. Cook for 5 to 10 minutes, stirring occasionally, or until the onion has softened and turned translucent. Add the sausage and cook for about 2 minutes, or until lightly browned. Add the stock and bring to a boil. Add the clams, cover, and cook for about 5 minutes, shaking the pan a few times, until the clams open.

Using a slotted spoon, place the clams in shallow serving bowls, discarding any clams that don't open. Stir the cilantro into the broth and pour it over the clams. Serve with lemon wedges.

NORA POUILLON'S SAKE-GLAZED BLACK COD

PREP TIME: 10 MINUTES | COOKING TIME: 20 MINUTES | SERVES 4

Nora Pouillon is a pioneer of the farm-to-table movement on the East Coast and the owner of Washington, DC's Nora, the first certified organic restaurant in the country. She helped start Washington's first producer-only farmers' market and is no stranger to ocean conservation, having participated in several seafood-focused environmental campaigns, including SeaWeb's Give Swordfish a Break.

"I first encountered this dish preparation at a tiny Japanese restaurant in Washington, DC, that was the favorite of Duncan, one of my chefs," Chef Pouillon explains. "He ordered the *omakase* dinner, which had five or six courses. One of them was this delicious caramelized perch, and I loved it so much, I asked the chef for the recipe, which he kindly supplied. I tweaked it for my restaurant, since the original asked for three times as much sugar. I like to serve my version using cod, with lots of stir-fried vegetables and a ginger-miso vinaigrette."

SAKE MARINADE (MAKES ABOUT ¾ CUP)
¼ cup sake or vodka

¼ cup mirin or sweet sherry

4 teaspoons sugar

⅓ cup white miso paste

BLACK COD
4 black cod (also called sablefish or Pacific cod) fillets (4 to 5 ounces each)

Combine the sake or vodka, mirin or sherry, and sugar in a small saucepan and cook over low heat for 2 to 3 minutes, or until the alcohol evaporates and the sugar dissolves. Whisk in the miso paste until smooth and let cool to room temperature.

Arrange the fish in 1 layer in a flat-bottomed glass or ceramic baking pan. Cover the fillets with a piece of moistened cheesecloth. Pour the sake marinade over the cheesecloth, spreading it out to cover the fish. (If you don't have cheesecloth, just spread the miso paste directly on the fish and scrape it off before cooking.) Marinate in the refrigerator for at least 1 hour, preferably overnight.

Preheat the broiler. Lift the cheesecloth with the miso paste off the fish. Place the fillets in a flameproof pan and broil 4" from the heat for 4 to 5 minutes, or until barely cooked through but caramelized on top.

Serve the fish with stir-fried vegetables, steamed rice, and lots of fresh cilantro.

★ TIP: *This sake-miso glaze recipe also works well with fillets of salmon, rockfish, barramundi, or halibut.*

MARIO BATALI'S JELLYFISH SALAD WITH
GOLDEN TOMATOES, OPAL BASIL, AND ARUGULA

PREP TIME: 10 MINUTES | SERVES 4

Mario Batali has built an empire of Italian cooking in the United States, and the inimitable chef, restaurateur, author, and TV show host is also an ocean conservationist. As he said in an interview with Nature.org (the Nature Conservancy's Web site): "Along with the atmosphere, the oceans are our largest resource; to squander any part of either unthinkingly is a fool's gain." And as some fishery experts will tell you, we have squandered the protein in our oceans to such a great extent that we may have to learn to love the taste of—you guessed it—jellyfish. Get started here with this tasty, whimsical recipe.

1 pound salted jellyfish, tentacles removed

1 pint yellow and red pear tomatoes, halved

10 opal (Thai) basil leaves, sliced thinly just before serving

1 bunch arugula, washed and spun dry, woody stems removed

2 tablespoons sherry vinegar

⅓ cup best-quality extra-virgin olive oil + additional for brushing bread slices

Kosher salt and freshly ground black pepper to taste

4 slices (each 1" thick) peasant bread, grilled or toasted

Soak the jellyfish for 30 minutes in a large bowl filled with cold water. Repeat until the water runs clear. Cut the body into thin slices and place in a large bowl.

Add the tomatoes, basil, and arugula to the bowl. Add the vinegar, ⅓ cup of the oil, and the salt and pepper and toss well to coat evenly. Divide the salad among 4 chilled dinner plates. Brush each toasted bread slice with olive oil. Serve immediately.

★ TIP: *Jellyfish can be found in most Asian markets. Be sure to wear gloves when handling. If you have trouble locating jellyfish, cooked squid also works wonderfully in this recipe.*

SAM HAZEN'S MIGNONETTE FOR OYSTERS

PREP TIME: 10 MINUTES | MAKES ABOUT 1½ CUPS (OR ENOUGH FOR AT LEAST 4 DOZEN OYSTERS)

During Chef Sam Hazen's career cooking in some of the world's best restaurants, including New York's Veritas, where he is now the executive chef and co-owner, he has learned an important lesson: Excellent ingredients sourced from local, trusted purveyors make all the difference. Likewise, one of the most important things about this recipe for mignonette, a vinegar-based sauce, is hidden between the lines: the source of your oysters. Buy them locally and in season if you can, and don't be afraid to ask a lot of questions to ensure quality and freshness.

½ cup red wine vinegar
½ cup finely diced tomatoes
1 tablespoon tomato juice
1 tablespoon finely grated lemon zest (about 2 lemons' worth)
1 tablespoon minced shallot
1½ teaspoons coarsely ground black pepper

In a small bowl, mix together the vinegar, tomatoes, tomato juice, lemon zest, shallot, and pepper.

Serve with the freshest available oysters on the half shell.

★ TIP: *Leftover mignonette can be refashioned into a delicious vinaigrette. Just blend with enough extra-virgin olive oil to soften the bite of the acidic sauce.*

CAT CORA'S HALIBUT WITH PEPITAS, CAPERS, CHERRY TOMATOES, AND BASIL

PREP TIME: 10 MINUTES | COOKING TIME: 15 MINUTES | SERVES 4

The first and only female Iron Chef on the Food Network, Cat Cora is a familiar force in the increasingly televised world of cooking. Raised in a small Greek community in Mississippi, Chef Cora often incorporates Mediterranean and Southern elements into her recipes. She says, "This is a restaurant-style dish that boasts a lot of flavor, thanks to a tomato-basil sauce with bursts of capers and the crunch of toasted pepitas." The earthiness of the pepitas nicely balances the salty capers and bright tomato.

2 tablespoons raw pepitas (pumpkin seeds)

4 halibut fillets (6 ounces each)

1 teaspoon salt + additional for seasoning

½ teaspoon freshly ground black pepper

2 tablespoons extra-virgin olive oil

1 cup dry white wine

½ cup halved cherry tomatoes (quartered if on the larger side)

2 tablespoons fresh lemon juice

3 tablespoons unsalted butter

1½ tablespoons capers, drained and rinsed

¼ cup torn fresh basil + a handful of small leaves for garnish

Place a small dry skillet over medium-low heat, add the pepitas, and toast them for 2 to 3 minutes, stirring and shaking the pan frequently, or until fragrant. Don't allow them to brown. As soon as you begin to smell their fragrance, remove the pan from the heat and transfer the pepitas to a large plate. Set aside to cool.

Preheat the oven to 400°F.

Sprinkle the halibut on both sides with salt and the pepper. Place a large skillet with an ovenproof handle over medium-high heat. Add the oil and heat until it begins to shimmer but not smoke. Place the fillets skin side up in the hot oil and cook for 2 minutes, or until the fish begins to brown lightly. Turn the fish over and add the wine, tomatoes, lemon juice, butter, and capers. Simmer for about 1 minute, or until the tomatoes release some of their juices.

Place the pan in the oven until the fish is opaque, firm to the touch, and cooked through. Start checking to see if the fish is done after 5 minutes (cooking time will depend on the thickness of the fillets). While the fish is cooking, occasionally baste it with some of the pan juices. Place the fish on warmed serving plates, reserving any accumulated juices in the pan. Add the torn basil, reserved pepitas, and the remaining 1 teaspoon salt to the pan juices and stir gently until the basil is slightly wilted. Spoon the mixture over the fish, garnish with the fresh basil leaves, and serve.

RICK BAYLESS'S WARM CLAM CEVICHE

PREP TIME: 25 MINUTES + SOAKING TIME
COOKING TIME: 10–15 MINUTES | SERVES 4 AS AN APPETIZER

Ceviche Tíbio de Almejas

A master of Mexican cuisine, Rick Bayless is best known for his Chicago restaurants Frontera Grill and Topolobampo, as well as his PBS show, *Mexico—One Plate at a Time*. Chef Bayless's restaurants are underpinned by a philosophy of environmental sustainability, reflected in his energy-efficient kitchens, local produce and seafood sourcing, composting, and more. This recipe showcases a delicious take on cooked ceviche rather than the more traditional version, in which the acid in the citrus "cooks" the fish.

2 pounds littleneck clams (about 24, depending on their size),
soaked for 20–30 minutes and drained

2 tablespoons unsalted butter, cut into small pieces

3 cloves garlic, peeled and finely chopped

Hot green chile pepper to taste (anything from ½ jalapeño
to 1 whole serrano, depending on how spicy you like it),
stemmed, seeded, and finely chopped

½ cup chicken broth

½ cup full-flavored light beer

¼ cup fresh lime juice

2 tablespoons chopped fresh cilantro

Salt to taste

Scrub the soaked clams well under cold running water. Place a deep 4-quart pot with a lid over medium heat and melt the butter. When the foam subsides, add the garlic and pepper and cook for about 1 minute, or until fragrant. Add the broth and beer. Increase the heat to high and bring to a boil. Add the clams and cover tightly. Cook for 3 to 5 minutes, or just until the clams open, discarding any that don't open. Divide the cooked clams among 4 serving bowls. Return the pot to high heat and continue to boil until the liquid is reduced by half. Remove from the heat and let the broth cool a couple of minutes. Add the lime juice and cilantro and season with salt, if necessary. Pour the broth over the clams and serve immediately.

SAM TALBOT'S THAI COCONUT MUSSELS

PREP TIME: 20 MINUTES | COOKING TIME: 10 MINUTES
SERVES 4–6 AS AN APPETIZER

Chef Sam Talbot grew up fishing for blue crab and flounder along the North Carolina coast, and he strives to serve seafood that's just as fresh, local, and sustainable. As the former executive chef at the Surf Lodge in Montauk, New York, and *Top Chef* Season 2 runner up told *Oceana* magazine: "I'm constantly looking for the most eco-responsible source. And usually when you find those people, their products, their fish, their vegetables, their honey, it's far superior to everything that's out there." Sweet, spicy, and layered with flavor, this dish can also be made with fresh littleneck clams if mussels aren't available. Serve with bowls of warm white sticky rice.

3 tablespoons extra-virgin olive oil

3 tablespoons peeled and finely chopped fresh ginger

4 large cloves garlic, finely chopped

1 shallot, finely chopped

2 tablespoons finely chopped lemongrass

2 tablespoons ground unsweetened coconut

1½ pounds mussels, debearded, scrubbed well, and washed clean

⅓ cup dry white wine

1 tablespoon soy sauce

1 teaspoon fish sauce

1 teaspoon sambal (chile-garlic paste)

¾ cup coconut milk

½ cup torn cilantro leaves, stems discarded

½ cup torn mint leaves, stems discarded

Juice and zest of 2 limes

Sea salt and freshly ground black pepper to taste

Warm the oil in a large saucepan over medium heat. Add the ginger, garlic, shallot, and lemongrass and cook for 2 to 3 minutes, or until fragrant. Add the coconut and stir frequently for 2 minutes, or until the shallot is translucent and the garlic and ginger have softened. Add the mussels and wine to the pan and increase the heat to medium-high. Cook for 1 minute. Add the soy sauce, fish sauce, and chile-garlic paste. Simmer for 1 minute. Stir in the coconut milk.

Cover the pan and steam the mussels for 2 to 4 minutes, or until they open. Discard any that don't open, and spoon the remaining mussels into a large warmed serving bowl. Stir the cilantro, mint, lime juice, and lime zest into the broth. Season with salt and pepper as needed. Serve immediately.

★ TIP: *Look for lemongrass, ground coconut, fish sauce, and sambal in your local Asian market.*

DAN BARBER'S VETA LA PALMA
MULLET ESCABÈCHE

PREP TIME: 10 MINUTES | COOKING TIME: 5 MINUTES
SERVES 4 AS A FIRST COURSE

Dan Barber isn't just a chef, he's a food philosopher. He believes we should know where the food on our plates comes from, and he's written about food and agricultural policy for the *New York Times*, the *Nation, Saveur,* and *Food & Wine*. And he practices what he preaches: One of his restaurants, Blue Hill at Stone Barns, is also a working farm and ranch. This recipe marries the quintessential Mediterranean dish escabèche (in which fish is marinated in a vinegar-based pickle) with mullet, in this case from Veta la Palma, a unique and sustainable aquaculture operation near Seville, Spain. The wetlands created by the fish farm are part of a biosphere reserve and provide a habitat for hundreds of species of European birds. Locally sourced mullet will work beautifully as well.

¾ cup white wine vinegar
¼ cup champagne vinegar
½ cup salt
¼ cup sugar
½ cup white wine
4 cups water
1 tablespoon coriander seeds
1 teaspoon black peppercorns
3 sprigs fresh thyme
1 side mullet, cleaned of pin bones and skin removed

In a medium saucepan, combine the vinegars, salt, sugar, wine, water, coriander seeds, peppercorns, and thyme. Place over medium-high heat and bring to a boil. Reduce the heat to low and let simmer for 2 minutes. Remove the pan from the heat and allow to cool to room temperature.

Cut the mullet into ½"-thick slices and arrange in a shallow glass or ceramic dish.

Heat just enough of the escabèche liquid to cover the slices and pour over the fish. Let the fish sit in the pickling liquid for about 1 minute. Remove the fish and place on a platter to serve. Discard the used pickling liquid. (Any unused escabèche liquid can be refrigerated for up to 1 month and is ideal for pickling vegetables as well.) Serve the fish with a lightly dressed green salad and grilled bread.

APRIL BLOOMFIELD'S OYSTER PAN ROAST
WITH TARRAGON TOASTS

PREP TIME: 15 MINUTES | COOKING TIME: 15–20 MINUTES | SERVES 4

Chef April Bloomfield hails from Birmingham, England, but you won't find any fish and chips on the menu at New York's critically acclaimed John Dory Oyster Bar, where she is chef and co-owner. Chef Bloomfield is known for her brazen cooking and her passion for seasonal, sustainable ingredients. This recipe, one of the John Dory's signature dishes, is a rich showcase for one of the ocean's most beloved bivalves.

4 tablespoons unsalted butter, softened

¼ cup fresh tarragon leaves, minced

1½ teaspoons fresh lemon juice

Kosher salt

1 tablespoon extra-virgin olive oil

½ small onion, minced

1 clove garlic, minced

¼ cup dry vermouth

¾ cup water

2 dozen oysters, such as Wellfleet, shucked,
¼ cup oyster liquor reserved

1 cup heavy cream

8 baguette slices, toasted

1 clove garlic

In a medium mixing bowl, mix the butter with the tarragon and ½ teaspoon of the lemon juice. Season with salt to taste. Set aside.

Heat the oil in a large sauté pan over medium heat. Add the onion and minced garlic. Cook for about 5 minutes, or until translucent and softened. Add the vermouth (carefully; it may flame) and increase the heat to medium-high. Simmer for 1 to 2 minutes, or until reduced by half. Add the water and reserved oyster liquor and simmer for 3 minutes. Add the cream and reduce the heat to medium. Simmer for about 5 minutes, or until the sauce thickens enough to coat a spoon. Remove from the heat. Add the remaining 1 teaspoon lemon juice. Season with salt.

Add the oyster meat to the sauce. Cook over medium-low heat for 2 minutes, or until cooked and warmed through.

While the oysters are cooking, assemble the toasts. Lightly rub the baguette slices with the whole garlic clove. Spread each toast with the reserved tarragon butter. Serve immediately.

TODD GRAY AND ELLEN KASSOFF GRAY'S
PICKLED HERRING IN CITRUS-DILL CRÈME FRAÎCHE

PREP TIME: 15 MINUTES | SERVES 6 AS AN APPETIZER

Chef Todd Gray and his wife, Ellen, co-owners of Washington, DC's Equinox, have been serving the flavors of the Mid-Atlantic for more than a decade, with a mission to use organic ingredients grown within 100 miles of the restaurant whenever possible. The duo are also supporters of Oceana's campaign to improve seafood traceability; they joined Oceana for a press conference on the subject in 2011. Chef Gray says that he didn't really eat herring until he met Ellen and her family. "On many an occasion, and especially at Sunday brunch, pickled or creamed herring would find its way onto the table right out of a jar," he says. "When I first tasted it, I found it oily, fishy, and tart from vinegar. So I decided to soften it up and give it a little bit of an upgrade by mixing it with some crème fraîche, citrus zest, and dill. I like to garnish this dish with skinless segments of orange, lemon, and lime—you may wish to do the same."

¾ cup crème fraîche or sour cream
½ teaspoon freshly grated orange zest
⅛ teaspoon freshly grated lemon zest
⅛ teaspoon freshly grated lime zest
1½ teaspoons chopped fresh dill
1 teaspoon rinsed, drained, and chopped capers
2 scallions, thinly sliced crosswise, including the green tops
1½ teaspoons freshly squeezed orange juice
1 jar (6 ounces) pickled herring, rinsed and drained well
Mixed greens and toasted sliced rye bread or crostini for serving

Place the crème fraîche or sour cream in a medium bowl. Stir in the orange, lemon, and lime zests and dill, capers, scallions, and orange juice, mixing until blended. Add the herring, stirring to coat the fillets evenly. Cover and refrigerate for at least 6 hours. Serve on a bed of mixed greens, accompanied by toasted rye bread or crostini.

MICHEL RICHARD'S FISH SOUP
WITH FLOUNDER, CRAYFISH, AND SQUID

PREP TIME: 15 MINUTES | COOKING TIME: 20 MINUTES | SERVES 4

Washington, DC's renowned chef-owner of Citronelle has a special relationship with seafood: He was born in Brittany on the northwest coast of France. Chef Michel Richard has been decorated with nearly every cooking award, including the James Beard award, and he is known for combining fresh California ingredients with traditional French cooking. This recipe, a simplified take on a traditional French bouillabaisse, is sure to satisfy.

4 ounces small boiling (waxy) potatoes, peeled and cut into ½" cubes
1 cup clam juice
½ cup heavy cream
¼ teaspoon saffron threads, crushed and dissolved in 2 tablespoons boiling water
Salt and freshly ground black pepper to taste
8 ounces medium crayfish, peeled and deveined
1 pound flounder fillet, trimmed and cut into ⅓" strips
8 ounces squid, cleaned and cut into ½" slices
1 small tomato, peeled, seeded, and diced

Set a steamer over a medium saucepan of water and steam the potatoes for about 10 minutes, or until tender and easily pierced with a knife. (This can be prepared ahead, covered, and set aside at room temperature.)

Place a large, heavy, nonstick skillet over medium heat and add the clam juice, cream, and saffron. Bring to a gentle simmer. Add the potatoes and season with salt and pepper. Bring the clam juice mixture to a boil, add the crayfish and flounder, reduce the heat, and simmer gently for about 2 minutes, or until the crayfish just start to turn pink, turning the seafood halfway through using tongs. Add the squid and cook for about 1 minute, or until opaque.

Using a slotted spoon, divide the seafood and potatoes among 4 warmed soup bowls. Quickly bring the clam juice mixture to a gentle simmer. Ladle over the seafood. Sprinkle with the tomato. Serve immediately.

HUGH FEARNLEY-WHITTINGSTALL'S
FISH-TOPPED PIZZA BIANCA

PREP TIME: 20 MINUTES ACTIVE + TIME TO ALLOW THE DOUGH TO RISE
COOKING TIME: 50 MINUTES | SERVES 4

Hugh Fearnley-Whittingstall is an author, a chef extraordinaire, and the founder of *Hugh's Fish Fight*, a television show and organization devoted to banning fish discards in the North Sea. He convinced more than 800,000 people to join him in calling for an end to policies that permit such excessive amounts of bycatch. This recipe, which is a delectable way to use up those tins of sardines you've been storing, is a keeper. Make the dough a day in advance, store it in the fridge, and you'll be rewarded with an easy after-work dinner.

DOUGH
1 package active dry yeast
⅔ cup warm water
1 cup all-purpose flour
1 cup bread flour
1 teaspoon salt
1 tablespoon olive oil

TOPPING
3 tablespoons olive oil + additional for baking
1½ pounds onions, very thinly sliced
Salt
A little flour or cornmeal for dusting
7 ounces canned sardines (or mackerel or other cooked oily or rich fish)
6–8 anchovy fillets, halved
A few tablespoons crème fraîche
Freshly ground black pepper
Chopped fresh parsley or chives, for serving

To make the dough: Dissolve the yeast in a little of the water. In a bowl, mix the all-purpose flour, bread flour, and salt. Stir in the yeast mixture, oil, and the remaining water. Mix until evenly combined.

Turn the dough out onto a lightly floured work surface and knead until smooth and stretchy. This combination of all-purpose flour and bread flour with a dash of olive oil produces a soft, elastic dough that is easy to roll.

Return the dough to the bowl, cover, and leave to rise in a warm place until doubled in size, or longer if you want. It will happily ferment all day; just punch it down every now and then to stop it from getting too big. It will even keep in the fridge for a couple of days.

To make the topping: Heat the oil in a large skillet, add the onions and a good pinch of salt, and cook gently over low heat, stirring occasionally, for about half an hour, or until soft and golden.

Preheat the oven to 500°F and put a baking sheet in it to heat.

Punch down the risen dough and cut it in half. Using a rolling pin, your hands, or both, roll and stretch one half into a very thin piece that will cover the baking sheet. Take the hot baking sheet from the oven, scatter it with a little flour or, even better, some cornmeal, and lay the dough on it.

Spread half of the onions over the dough and scatter with half of the fish, including the anchovies. Add a few dollops of crème fraîche. Season with some salt and pepper, add a trickle of olive oil, and place in the oven. Bake for 10 to 12 minutes, or until the base is crisp and golden brown at the edges.

While it's cooking, roll out the remaining dough and top it, so it is ready to go in the oven as soon as the first pizza is cooked. Slice and serve hot, sprinkled with parsley or chives.

LAURENCE JOSSEL'S LITTLE FRIED FISH
WITH LEMON-DILL AIOLI

PREP TIME: 25 MINUTES | COOKING TIME: 10 MINUTES | SERVES 4

Laurence Jossel is currently the chef at San Francisco's critically acclaimed Nopa, but his experience at Kokkari, cooking Mediterranean food with a solidly Greek bent, is exemplified by this dish. Versions of it are served all over Greece in the same way Americans enjoy french fries: as delicious, simple finger food. Crunchy, hot, and lip-smackingly succulent, little fried fish drizzled with garlic-infused lemony aioli is one of life's great pleasures and an ideal way to enjoy fish that are often overlooked.

LEMON-DILL AIOLI

1 small clove garlic, grated with a microplane

1 tablespoon freshly squeezed lemon juice

Zest of 1 lemon

1 egg yolk

½ cup olive oil

½ cup neutral vegetable oil

2 tablespoons chopped fresh dill

LITTLE FRIED FISH

40 whole anchovies or smelt or 20 sardine fillets

1 quart buttermilk

2 cups all-purpose flour

½ cup cornmeal

5 tablespoons salt

1 tablespoon ground red pepper

Vegetable oil for frying

Lemon wedges for serving

Dill sprigs for serving

To make the aioli: In a blender or food processor, combine the garlic, lemon juice, lemon zest, and egg yolk. Process until the ingredients are combined. With the motor running, add the olive and vegetable oils in a very slow, steady stream. Allow the oil to mix in fully before adding more, without letting pools of oil form. If the mixture becomes very thick, sprinkle in a small amount of cold water to keep it from breaking. Place the aioli in a small bowl. Fold in the dill by hand. Cover and refrigerate until the fish are ready.

To make the fish: Clean the anchovies or smelt by holding each with its belly facing up and then pinching and pulling out the gills. Push your finger into the opening behind the gills and run it backward into the belly to open and remove the stomach. It is okay if the head comes off, but it looks nice if it stays on. Rinse the fish in water, place them in a large bowl, and cover with the buttermilk. If you're using cleaned sardine fillets, simply place them in the buttermilk.

In a medium bowl, whisk together the flour, cornmeal, salt, and pepper.

In a large, straight-sided skillet, add enough vegetable oil to cover a single layer of fish and heat to 375°F. Line a baking sheet with paper towels or a rack. Set aside.

Dredge the fish with the flour-cornmeal mixture and set on a large plate. Working in batches, fry a single layer of fish for about 3 minutes, or until it has a nice golden color.

Drain the fish on the prepared baking sheet. While still warm, serve with the aioli. Garnish with lemon wedges and dill sprigs.

JOSÉ ANDRÉS'S CLEMENTINES WITH CHINCHÓN, ANCHOVIES, AND BLACK OLIVES

PREP TIME: 15 MINUTES | SERVES 4

Born in Asturias, Spain, Jose Andrés grew up surrounded by stellar seafood. Here, the chef who introduced the concept of "small-plate dining" to a grateful American public combines Mediterranean clementines with the earthy, anise-infused flavor of Spanish Chinchón from the heart of Spain and marries them to succulent anchovies and black olives. Geographically, these ingredients would never be found together, but in this salad, they exist in perfect harmony. Chef Andrés advises, "If you can't find Chinchón, try another anise-flavored liqueur, like Marie Brizard. And if you don't have chervil for the garnish, use a little flat-leaf parsley."

DRESSING

½ cup Spanish extra-virgin olive oil

2 tablespoons sherry vinegar

2 tablespoons Chinchón or other anise-flavored liqueur

Zest and juice of 1 clementine

1 small shallot, thinly sliced

Sea salt to taste

SALAD

½ head romaine lettuce (tough outer leaves discarded), washed and spun dry

½ head Bibb lettuce (tough outer leaves discarded), washed and spun dry

½ head red leaf lettuce (tough outer leaves discarded), washed and spun dry

4 clementines, peeled and sliced into ¼" rounds

12 empeltre olives or other good-quality black olives

8 oil-packed anchovy fillets

Sea salt to taste

Fresh chervil or watercress for garnish

To make the dressing: In a medium bowl, whisk together the oil, vinegar, liqueur, clementine juice and zest, and shallot in a medium bowl. Season with salt and set aside.

To make the salad: Separate the heads of lettuce into individual leaves and divide among 4 plates or bowls. Divide the clementines, olives, and anchovies among the salads. Drizzle each salad with some of the dressing. Season with salt. Garnish with chervil or watercress.

CARLA HALL'S SQUID 'N' GRITS

PREP TIME: 15 MINUTES | COOKING TIME: 20 MINUTES | SERVES 4 TO 6

Carla Hall is best known as a competitor on Bravo's *Top Chef*, where she won over audiences and judges with her sunny personality and Southern-inspired food. A native of Nashville, Tennessee, she started her career as an accountant, but was inspired to change careers after a delicious trip through Europe. Her approach to cooking blends her classic French training and Southern heritage, reflected in this tasty take on shrimp and grits.

1½ cups whole milk
1½ cups vegetable broth or clam juice
Salt and freshly ground black pepper to taste
1½ cups stone-ground grits
3 tablespoons butter
4 strips thick-cut bacon (about 4 ounces), finely diced
1 small onion, finely diced
2 celery stalks, finely diced
1 leek, white part only, finely diced
2 garlic cloves, minced
4 sprigs fresh thyme
Zest of 1 lemon
Chile flakes to taste
2 tablespoons all-purpose flour
½ cup fish broth or clam juice
2 bay leaves
1 tablespoon canola oil, or more as needed
2 pounds cleaned calamari, bodies sliced into ¼" rings,
tentacles cut in half if large
½ cup scallions, thinly sliced, including the green tops
¼ cup flat-leaf (Italian) parsley, chopped

In a 5-quart saucepan over medium heat, combine milk and vegetable broth or clam juice. Season with salt. Whisking constantly, slowly pour the grits into the liquid. Bring the mixture to a boil, whisking constantly, and reduce the heat to low. Maintain a simmer and stir once every few minutes until the grits are soft and creamy, about 20 minutes. If the grits begin to look dry, add more broth as necessary, in tablespoons. Stir in the butter, and season to taste with salt and pepper. Remove from heat and set aside.

Place the bacon in a large skillet over medium heat and cook until crisp, turning frequently. Using a slotted spoon, remove bacon from skillet and set aside on a small plate. Reserve the bacon fat in the pan. Add the onion, celery, leek, and garlic to the bacon fat. Cook until softened and the onion has turned translucent and begun to soften, about 5 to 7 minutes. Add the thyme, lemon zest, and chile flakes. Sprinkle the flour over the vegetables and stir to incorporate. Cook for another 2 to 3 minutes. Add fish broth or clam juice and bay leaves. Bring to a boil, reduce the heat to medium, and let simmer until the mixture has thickened, about 5 minutes. Season to taste with salt and pepper. Keep warm over low heat.

Place a large skillet over medium-high heat. Add enough canola oil to coat the bottom of the pan. Working in batches, sauté a single layer of calamari until opaque and cooked through, about 1 to 2 minutes. Add the cooked calamari to the vegetable mixture. Continue cooking any remaining calamari. Once all the calamari is cooked, garnish the vegetable-calamari mixture with scallions and parsley. Serve immediately over the grits.

GASTÓN ACURIO'S CAUSA LIMEÑA WITH ANCHOVIES

PREP TIME: 30 MINUTES | COOKING TIME: 45 MINUTES | SERVES 4

With 32 restaurants in 12 countries, prolific Peruvian chef Gastón Acurio is one of Latin America's most esteemed chefs. Acurio and his wife operate restaurants around the world, including the popular upscale global chain La Mar Cebichería, which opened its doors in New York City in 2011. Acurio is dedicated to seafood sustainability, and the restaurant has partnered with the Monterey Bay Aquarium's Seafood Watch program, pledging to source only sustainable seafood. This traditional Peruvian potato dish benefits from the addition of anchovies, which are caught in abundance in Peru, but are too often ground up to feed livestock instead of, well, you. Be sure that the anchovies are fresh and preferably caught the day you plan to cook this dish.

1½ pounds Yukon gold potatoes, scrubbed well (try to select potatoes that are equal in size, so they cook in roughly the same time)

¼ cup plus 2 tablespoons aji amarillo paste, divided

2 tablespoons canola oil

Juice of 3 limes, divided

Salt and freshly ground black pepper to taste

For the salsa:

1 small red onion, peeled, sliced in half, and julienned

1 aji limo, finely chopped

2 tablespoons finely chopped flat-leaf (Italian) parsley

½ cup mayonnaise

2 avocados, cubed

Vegetable oil for frying

16 whole anchovies (32 fillets), cleaned, with skin and bones removed

1 cup all-purpose flour

¼ cup quartered cherry tomatoes (about 6 to 8 tomatoes)

Prepare the potato cakes: Place the potatoes in a medium saucepan and cover with cold water. Place over medium-high heat and bring to a boil. Cook the potatoes until tender when pierced with a knife. While the potatoes are still warm, peel them. Pass the peeled potatoes through a potato ricer set over a large bowl. Add ¼ cup aji amarillo paste, canola oil, and juice of 1 lime to the potatoes. Season to taste with salt and pepper. Set aside and keep warm.

Prepare the salsa criolla: Place the red onion, aji limo, and parsley in a small bowl. Add juice of 2 limes and 1 tablespoon water. Mix together. Season to taste with salt and pepper. Set aside.

In a small bowl, mix the mayonnaise with 2 tablespoons aji amarillo paste. Keep refrigerated until ready to serve. In a small bowl, mash the avocado with a fork. Season to taste with salt and pepper. Set aside.

Pour oil into a heavy saucepan to a depth of 3 inches. Place the saucepan over medium-high heat. Line a baking sheet with paper towels. While the oil is heating, season the anchovies with salt and pepper. Place the flour in a shallow bowl. Lightly coat the anchovies with flour. Shake off any excess flour.

When the oil has reached 360°F, fry the anchovies in batches, until they are golden brown and crispy, about 1 to 2 minutes. Using a slotted spoon, transfer the anchovies to the prepared baking sheet and keep warm in the oven. Repeat with the remaining anchovies.

Roll the *causa* (potato) dough in a log. Cut into four equal pieces. Place each piece of causa dough on 4 serving plates. Top each with the mashed avocado, cherry tomatoes, and fried anchovies. Drizzle the salsa criolla over the top. Decorate the plate with the aji amarillo cream. Serve immediately.

★ TIP: *Aji limo and aji amarillo are available at most international grocery stores.*

HAJIME SATO'S SABA (MACKEREL) WITH DILL

PREP TIME: 1 HOUR | COOKING TIME: 5 MINUTES | SERVES 2

Hajime Sato, chef/owner of Seattle's Mashiko sushi bar, is outspoken about seafood sustainability. So much so that in 2009, 15 years after opening the restaurant, he decided to make Mashiko Seattle's first fully sustainable sushi bar, which meant nixing some of the sushi world's most beloved but overfished species, like bluefin tuna and eel. Instead of big ocean predators like bluefin, Mashiko serves up small fish like herring, sardines, and, yes, mackerel, or "saba" in Japanese. This simple preparation of lightly fried mackerel with dill and sake will make you forget all about the big fish.

2 (4- to 6-ounce) mackerel fillets, cleaned and pin bones removed
Salt as needed
2 tablespoons chopped fresh dill, plus more for garnish
½ cup all-purpose flour
2 tablespoons grapeseed oil or other neutral vegetable oil
2 tablespoons sake
Lemon wedges and grated daikon radish for serving

Place the mackerel fillets on a large plate. Sprinkle on both sides with salt. Cover loosely with plastic wrap, place in refrigerator. Let sit for 1 hour. Pat mackerel dry with a paper towel. Press the 2 tablespoons dill into the skin side of the mackerel.

Place the flour in a shallow bowl. Coat the mackerel on both sides with flour.

Over medium-high heat, warm the oil in a nonstick frying pan large enough to hold the mackerel in a single layer. Place the mackerel in the pan, skin side down. Cook until light golden brown, 1 to 2 minutes. Gently turn the mackerel over and cook another 1 to 2 minutes. Pour the sake over the mackerel and put on the lid until the sake evaporates, about 15 to 20 seconds. Serve on warmed plates with additional chopped dill, lemon wedges, and grated daikon.

LIDIA BASTIANICH'S GLOUCESTER BAKED HALIBUT

PREP TIME: 10 MINUTES | COOKING TIME: 20 MINUTES | SERVES 4

Lidia Bastianich is a star of Italian-American cuisine. She is the chef-owner of several acclaimed Italian restaurants in New York, including Felidia, Becco, Esca, and Del Posto, and has appeared in a number of television cooking series, including *Lidia's Italy*. With her son, Joe Bastianich, Mario Batali, and Oscar Farinetti, she opened Eataly, the largest artisanal Italian food and wine marketplace in New York City, in 2010. Her baked halibut is great for a beginner to cooking seafood; just be sure to choose Pacific halibut, since the fish stocks are in much better shape than Atlantic halibut. When you serve it up you might say, as Lidia does, *Tutti a tavola a mangiare*—Come to the table and eat!

3 garlic cloves, peeled

¼ teaspoon kosher salt, plus more for seasoning the fish

½ cup extra-virgin olive oil

1 cup crushed saltine crackers

1 tablespoon chopped fresh thyme

1 cup grated white cheddar cheese

4 (6- to 8-ounce) skinless halibut fillets

Preheat oven to 400°F. Add the garlic and ¼ teaspoon salt to the olive oil in its measuring cup. Let steep for 30 minutes (or longer, time permitting). Place the saltines and thyme in a medium bowl. Pour in ¼ cup of the infused olive oil, leaving the sliced garlic behind in the cup. Coat all of the crumbs with olive oil. Fold in the grated cheese.

Spread the crumbs on a plate. Brush each fish fillet all over with the remaining oil and season with salt. Use any oil left in the cup to grease a rimmed baking sheet. Roll the fish on all sides in the crumbs, pressing to coat well, trying to use all of the crumbs. Put the fish on a baking sheet, leaving an inch or so between each fillet.

Bake the fish until the crumbs are crisp and golden and the fish is cooked through, about 20 minutes. Serve immediately.

EMERIL LAGASSE'S SPANISH MACKEREL WITH SPINACH & CHICKPEA SALAD & TOMATO VINAIGRETTE

PREP TIME: 15 MINUTES | COOK TIME: 25 MINUTES | SERVES 6

An instantly recognizable food personality, Emeril Lagasse has hosted more than 2,000 shows on the Food Network, and is the food correspondent for ABC's *Good Morning America*. Emeril owns 13 restaurants and has published 17 cookbooks, mostly devoted to Creole and Cajun cooking, but he got his start making bread and pastry at a Portuguese bakery in his Fall River, Massachusetts, neighborhood. This recipe is rich and packed with omega-3s, but make sure to get the freshest mackerel you can—look for firm, shiny bodies and clear, bright eyes.

2¼ cups tomato sauce, slightly warmed and divided

¼ cup sherry vinegar

2 small garlic cloves, minced

Salt and freshly ground black pepper to taste

¾ cup plus 1 tablespoon olive oil, divided

6 cups (6 ounces) baby spinach

¼ cup thinly sliced red onion

¼ cup julienned red bell pepper

¼ cup pitted and sliced black olives

6 (4- to 6-ounce) Spanish mackerel fillets, cleaned and pin bones removed

2 cups cooked (15-ounce can) chickpeas, rinsed and drained

2 tablespoons chopped fresh parsley

Preheat oven to 375°F.

Make the vinaigrette: Place ¼ cup tomato sauce, vinegar, garlic, and one pinch each of salt and pepper in a small mixing bowl. Whisk until combined. Continue to whisk vigorously while slowly adding ¾ cup olive oil in a slow, steady stream. Season with additional salt and pepper as needed. Set aside.

Place spinach, red onion, bell pepper, black olives, and one pinch each of salt and pepper in a large mixing bowl. Set aside.

Brush both sides of mackerel with 1 tablespoon olive oil. Season to taste with salt and pepper. In a roasting pan, add 2 cups warmed tomato sauce. Place mackerel in the tomato sauce and place in the oven. Cook until mackerel is firm to the touch, opaque and just cooked through, 8 to 10 minutes. Remove from oven and place the mackerel on a plate.

While the mackerel is cooking, place a large sauté pan over medium heat. Add ¼ cup tomato vinaigrette, chickpeas, and one pinch each of salt and pepper. Cook until warmed through, 1 to 2 minutes. Keep warm over low heat.

Pour chickpeas over spinach mixture. Using a spatula, scrape any residual vinaigrette from the pan into the bowl. Drizzle another 1 to 2 tablespoons of vinaigrette over salad, as needed. Gently toss the salad, making sure all spinach leaves are coated well.

Divide spinach and chickpea salad among 6 serving plates. Top with mackerel fillets, drizzle with extra tomato vinaigrette as desired, and garnish with parsley. Serve immediately.

SEAN BROCK'S CAROLINA TRIGGERFISH
WITH HEIRLOOM BEANS AND CORN

PREP TIME: 25 MINUTES, PLUS TIME FOR SOAKING BEANS |
COOKING TIME: 1 HOUR, 20 MINUTES | SERVES 4

In just a few years, Sean Brock has rocketed to a top spot among American chefs. His two restaurants in Charleston, South Carolina, McCrady's and Husk, have helped revitalize high-end cuisine in the Southeast. A young chef and a Virginia native, Brock hews closely to Southern tradition with his cooking. He sources as many ingredients as possible from the South, including the seafood. If you live in or visit the Southern coast, be sure to try triggerfish—you won't be disappointed by this firm, white-fleshed fish.

4 to 5 cups chicken, vegetable, or pork broth

1 medium onion, diced

1 large carrot, diced

2 celery stalks, diced

2 garlic cloves, thinly sliced

1 bay leaf

3 sprigs fresh thyme

½ jalapeño, stemmed, seeded if desired, and chopped

1 cup medium-size dried heirloom beans (such as flageolet), soaked in water and refrigerated overnight, drained the following day

6 ears corn, shucked and cleaned

4 tablespoons unsalted butter, divided

1 small bunch chives, minced

4 (7-ounce) triggerfish fillets

Salt, cayenne pepper, and freshly ground black pepper to taste

Canola oil as needed

Juice of 1 lemon

1 small bunch scallions, thinly sliced, including the green tops

In a large saucepan, bring 4 cups broth to a simmer and add onion, carrot, celery, garlic, bay leaf, thyme, jalapeño, and drained, soaked beans. Cook over low heat, partially covered, until the beans are tender, about 1 hour. Season to taste with salt. (Beans can be cooked 1 day in advance, drained, and refrigerated. Warm gently before serving.)

Cut the corn kernels away from the cob and place in a large sauté pan, making sure to collect any accumulated liquid from the corn. Place over medium heat and cook gently to soften the corn kernels and reduce the corn liquid. If there is not any or enough corn liquid, add ⅓ cup water to the pan to cook the corn. Once the corn juice thickens, stir in 2 tablespoons butter. Add the cooked beans to the corn, along with a small amount of the bean-cooking liquid. Stir in the chives. Season to taste with salt and pepper.

Season the triggerfish with salt and cayenne pepper. Place a medium sauté pan over medium-high heat. Add enough canola oil to cover the bottom of the pan. When oil is shimmering, place the fillets into the pan. Cook until golden brown, 2 to 3 minutes each side. Top each fillet with ½ tablespoon butter and a drizzle of lemon juice. Serve each triggerfish fillet with some of the bean-corn mixture. Garnish with scallions and serve.

RECIPE CREDITS

"SMOKED SARDINES WITH HEIRLOOM TOMATOES AND HERBS" from *For Cod and Country* © 2011 by Barton Seaver, printed courtesy of Sterling Epicure. All rights reserved.

"CLAMS WITH SPICY SAUSAGE" printed courtesy of Eric Ripert.

"SAKE-GLAZED BLACK COD" printed courtesy of Nora Pouillon.

"JELLYFISH SALAD WITH GOLDEN TOMATOES, OPAL BASIL, AND ARUGULA" from *The Babbo Cookbook* by Mario Batali, © 2002 by Mario Batali. Photographs © 2002 by Christopher Hirsheimer. Used by permission of Clarkson Potter/ Publishers, an imprint of the Crown Publishing Group, a division of Random House, Inc. Any third party use of this material, outside of this publication, is prohibited. Interested parties must apply directly to Random House, Inc. for permission.

"MIGNONETTE FOR OYSTERS" printed courtesy of Sam Hazen.

"HALIBUT WITH PEPITAS, CAPERS, CHERRY TOMATOES, AND BASIL" from *Cat Cora's Classics with a Twist* by Cat Cora, © 2010 by Cat Cora. Used by permission of Houghton Mifflin Harcourt Publishing Company. All rights reserved.

"WARM CLAM CEVICHE" printed courtesy of Rick Bayless.

"THAI COCONUT MUSSELS" printed courtesy of Sam Talbot.

"VETA LA PALMA MULLET ESCABÈCHE" printed courtesy of Dan Barber.

"OYSTER PAN ROAST WITH TARRAGON TOASTS" printed courtesy of April Bloomfield.

"PICKLED HERRING IN CITRUS-DILL CRÈME FRAÎCHE" from *The New Jewish Table* © 2013 by Todd Gray and Ellen Kassoff Gray. Reprinted by permission of St. Martin's Press. All rights reserved.

"FISH SOUP WITH FLOUNDER, CRAYFISH, AND SQUID" printed courtesy of Michel Richard.

"FISH-TOPPED PIZZA BIANCA" from *The River Cottage Fish Book*, text © 2007 by Hugh Fearnley-Whittingstall and Nick Fisher. Photographs © 2007 by Simon Wheeler. Additional photography © 2007 by Paul Quagliana, Marie Derome, and other contributors (see page 606). Used by permission of Ten Speed Press, an imprint of the Crown Publishing Group, a division of Random House, Inc. Any third party use of this material, outside of this publication, is prohibited. Interested parties must apply directly to Random House, Inc. for permission.

"LITTLE FRIED FISH WITH LEMON-DILL AIOLI" printed courtesy of Laurence Jossel.

"CLEMENTINES WITH CHINCHÓN, ANCHOVIES, AND BLACK OLIVES" from *Made in Spain* © 2008 by José Andrés. Photographs © 2008 by Thomas Schauer. Used by permission of Clarkson Potter/Publishers,

an imprint of the Crown Publishing Group, a division of Random House, Inc. Any third party use of this material, outside of this publication, is prohibited. Interested parties must apply directly to Random House, Inc. for permission.

"SQUID 'N' GRITS" printed courtesy of Carla Hall.

"CAUSA LIMEÑA WITH ANCHOVIES" printed courtesy of Gastón Acurio.

"SABA (MACKEREL) WITH DILL" printed courtesy of Hajime Sato.

"GLOUCESTER BAKED HALIBUT" from *Lidia's Italy in America* by Lidia Matticchio Bastianich and Tanya Bastianich Manuali, © 2011 by Tutti a Tavola, LLC. Photographs © 2011 by Christopher Hirsheimer. Used by permission of Alfred A. Knopf, a division of Random House, Inc. Any third party use of this material, outside of this publication, is prohibited. Interested parties must apply directly to Random House, Inc. for permission. More information on Lidia's recipes may be found at www.lidiasitaly.com

"SPANISH MACKEREL WITH SPINACH & CHICKPEA SALAD & TOMATO VINAIGRETTE" by Emeril Lagasse originally appeared on *Emeril's Table* on Hallmark Channel, courtesy of Martha Stewart Living Omnimedia, Inc.

"CAROLINA TRIGGERFISH WITH HEIRLOOM BEANS AND CORN" printed courtesy of Sean Brock.

AN EASY GUIDE FOR EATING SEAFOOD RESPONSIBLY

WE WANT TO ENSURE that we can all enjoy eating seafood for generations to come. As we've said over and over again, fish is the "perfect protein" because it's better for the planet and better for you. However, we recognize that it can get confusing when you go to your local fish market or restaurant and try to find the "right" seafood, so here are some guidelines we hope will help.

First, pat yourself on the back. The mere fact that you are reading this book and engaging with ocean issues is significant. Securing better policies for the oceans is the best way to make sure our seas are full of fish for generations to come; understanding this concept makes you a responsible seafood eater before you even lift a fork.

Second, pick up one of the great responsible seafood guides available, such as the Monterey Bay Aquarium's Seafood Watch Guide, which you can find at seafoodwatch.com and is also available as a wallet-sized booklet and as an app. All of the guides can serve as informative and helpful chaperones for your next seafood purchase. They are necessarily specific as each fishery in the world is subject to different management, abundance, and habitat conditions and can be caught with gear types that range from selective to ecologically catastrophic.

Third, do not worry about being perfect—just try to be good. Too often, the teenager working the summer job behind your local fish counter or as a server at your favorite seafood restaurant may not know whether the fillet of "cod" you want to eat was caught off of Iceland with a hook and line or on Georges Bank with bottom trawl gear (or is in fact cod at all). So try to base your seafood decisions on the following simple rules and use the seafood guides when you can. Finally, don't kick yourself if you make a mistake. It's a long-term effort and every little bit counts.

Rules for responsible seafood eaters

• FOLLOW THE MAXIM: "EAT WILD SEAFOOD. NOT TOO MUCH OF THE BIG FISH. MOSTLY LOCAL." •

In our homage to Michael Pollan's memorable *Food Rules,* we advise that you generally try to eat wild seafood. The big exception to this rule is that we encourage you to enjoy farmed shellfish. Otherwise, eat lots of small fish like mackerel, eat big fish rarely, and, when you do, try to stick to the safe bets— like wild salmon—and otherwise use the seafood guides to find your best choice. When you can't find anything on the menu that's in the seafood guide, as sometimes happens, opt for local fish—there are often great options (such as bluefish in New York, for example).

• ENJOY LOTS OF SHELLFISH •

Shellfish is generally a good bet for the responsible seafood eater. The term shellfish can refer to a wide variety of marine invertebrates that often have little in common biologically. Plus, the farming of bivalve mollusks like oysters and mussels can actually improve the environment around them.

First, the really good news: Lovers of oysters, mussels, and clams can confidently slurp away. As filter feeders, these animals, farmed or wild, actually improve water quality and can be a panacea for estuaries and coastal areas degraded from nutrient pollution. A single oyster is capable of filtering 20 gallons of water a day and can form reefs that protect against storm surges. There is virtually no downside in indulging your love of these bivalves.

Things get a bit trickier with scallops. Diver scallops and bay scallops can be a wonderful choice—they are caught with minimal impact to the environment. Plain old "scallops" are caught with dredges that can destroy seafloor habitat (and often treated with chemicals that make them last longer and retain lots of water). They are not so great.

Enjoy lobsters and crabs! These crustaceans are closer relatives to butterflies than bivalves, but they often are grouped into the shellfish category (so we are doing the same here). Lobsters and crabs are caught in pots, a method that is associated with little or no bycatch. These are generally sustainable, delicious choices. So you can enjoy your lobster ravioli or she-crab soup in good conscience.

Now, for the bad news: If you want to be a responsible seafood eater, you are going to have to eat less shrimp. Shrimp is one of the most—if not the most—damaging fisheries around. The news is grim for this cocktail-party and bayou-barbecue staple. Even in the highly regulated United States, 76 percent of the marine life that shrimp trawlers haul up isn't shrimp at all but rather species like shark, red snapper, and almost 9,000 endangered sea turtles each year. Most of the shrimp you will likely buy or eat—9 out of 10 consumed in the United States—is farmed and imported. And the majority of farmed shrimp comes at a heavy cost to the environment, with pristine tropical mangroves destroyed to make way for industrial farms that spread pollution and disease. These farms not only degrade the environment but also the prospects for artisanal fishermen, who watch as habitat crucial to their local fisheries is demolished. If you are determined, you can find cold-water shrimp in the supermarket from the North Atlantic and the Pacific that are fished more sustainably. But you are going to need to use your guide to find the responsible stuff.

• STAY SMALL (EAT THE LITTLE FISH) •

As we hope has been made abundantly clear, it is difficult to beat eating whole forage fish like anchovies, herring, and sardines, both environmentally and nutritionally. We enthusiastically recommend eating any of the small stuff fresh whenever you can find it. A grilled fresh sardine or anchovy is many a chef's (and our) favorite seafood delight. Just see all the recipes we've been able to compile from top chefs in this book. When you can't find fresh versions of the smallest fish, ask your fishmonger or server for other small(er) species found on the seafood guides, like tilapia. You should be able to procure canned or bottled sardines, anchovies, and herring at your local market; they are delicious by themselves or as a tasty addition to other dishes. Remember, small fish are largely free from the toxins that accumulate in larger fish and are generally caught without using destructive bottom trawling methods that can destroy centuries-old seafloor communities. Little fish could feed hundreds of millions of people sustainably and healthily if managed wisely.

By opting for small fish, you vote with your mouth. Every little fish you eat is one that is not ground up and inefficiently fed to livestock or to salmon and tuna in farms that despoil surrounding waters. If more of us ask for these delicious fresh small fish, we'll be more likely to find it at our seafood counters and on local menus. Let's make sure that seafood lovers in Europe and elsewhere aren't the only ones who get to regularly enjoy freshly grilled sardines.

• EAT BIG MINDFULLY AND RARELY •

It's an unfortunate fact that if you want to be a responsible fish eater, you have to be mindful and limited in your consumption of big fish. By big fish we mean tuna, swordfish, and the other large fish at the upper reaches of the ocean food web. After decades of plunder, there simply aren't that many big fish left in the ocean. Also, big fish are generally a less healthy choice: Mercury and other pollutants accumulate in large predators like tuna. So, as a responsible fish eater, you should treat big fish as an occasional treat.

The appeal of carving into a hefty swordfish steak or filleting a marbled salmon is undeniable. The good news is that there are ways to be a responsible consumer when it comes to enjoying larger fish. First, use the guides—they are incredibly helpful in picking a good choice. And focus on the safe bets, like wild Alaskan salmon. Always avoid certain fish, such as bluefin tuna or Chilean sea bass, which are both in real jeopardy. Study the guides' green and red lists for other options to look for and avoid.

With big fish, try to eat local. The United States has some of the best-regulated wild fisheries in the world, especially in Alaska, where stocks of wild salmon, halibut, and other fish are all generally well managed. While these fisheries are far from perfect (and Oceana and our allies are fighting to improve them), there is a better chance that if you are eating a wild American fish, it is being managed with science-based quotas and the fishermen catching it are not destroying habitats in pursuit of your fish.

Also, stay wild with big fish. As previously discussed, there are a number of problems with fish farms that raise salmon and other carnivorous big fish, from the heavy use of antibiotics to the fact that they are contributing to over-fishing by consuming ground wild fish.

• GO WILD, NOT ORGANIC •

You might have seen "organic" fish at your local restaurant. The truth is that there is no such thing as an organic wild fish given that the "organic" label refers to how the fish is raised, as is the case when you buy organic milk, apples, or other food. There are some organic fish farms outside of the United States, but none of them is yet certified by US authorities. You are going to need to do extensive research into a specific farm's practices before you can be confident about eating organic as a responsible seafood eater. Instead, we recommend that you stick with the wild and local labels to guide your responsible seafood choices.

• YOU DON'T HAVE TO BECOME A GEARHEAD •

You probably have seen "how your fish was caught" descriptions on your menu that specify the type of fishing gear used. Some of the better methods can mitigate bycatch and habitat destruction. For example, pole, troll, hook-and-line, and harpoon-caught are good terms to look for as a responsible eater (particularly for bigger fish). If possible, stay away from trawls, driftnets, gill-nets, and longlines, which are notorious for catching unwanted species and damaging habitat. Unfortunately, it can often be nearly impossible to find out how your fish was caught. So you don't need to become a total gearhead to be a responsible seafood eater; if you can elicit any useful information about the fishing method, wonderful—but again, you just need to aim for good, not perfect.

• GET YOUR OMEGA-3S FROM A TIN OR SMALL FISH (RATHER THAN A SUPPLEMENT) •

It is often impossible to determine what type of fish was used to make your fish oil capsules. So we recommend that as a responsible seafood eater, you should get your omega-3s from a tin of sardines or anchovies (or, even better, fresh forage fish) rather than from a supplement. If you have to take fish oil supplements, try to find a brand that actually indicates on the label what fish is being used in the supplement, and choose a source that is green-listed in the seafood guides (or in our rules)—for example, fish oils that are derived from "100% Wild Alaska Salmon."

ACKNOWLEDGMENTS

MANY PEOPLE CONTRIBUTED to the creation of this book. We are especially grateful to the communities and companies that allowed us to glimpse real-world achievements in sustainable seafood. This includes Paul Butler, Stuart Green, Fel Cadiz, Lito Mancao, Cherry Ravelo, Vincent Dueñas, Marybeth Rita, Julie McCord, and the people of Lanuza Bay, Philippines, who could not have been more welcoming.

In Bethel, Alaska, we were thankful for the generous time and attention of Myron Naneng and Timothy Andrew of the Association of Village Council Presidents. Chad Carroll was a kind and knowledgeable host as he revealed the inner workings of a family-owned oyster and clam farm in southern Virginia. Kevin McDade, Sarah Haser, Carrie Brownstein, and the staff of Whole Foods welcomed our probing questions. Paul Willis of Niman Ranch, Captain Bouncer Smith, Sheila Bowman of the Monterey Bay Aquarium, Bob Rheault, and Roz Naylor were invaluable. Joyce Nettleton guided us toward some understanding of the knotty details of omega-3 biochemistry.

We are incredibly grateful to the chefs who shared their culinary visions for a sustainable seafood future, including April Bloomfield, Barton Seaver, Carla Hall, Cat Cora, Dan Barber, Ellen Kassoff Gray, Emeril Lagasse, Eric Ripert, Gastón Acurio, Hajime Sato, Hugh Fearnley-Whittingstall, José Andrés, Lawrence Jossel, Lidia Bastianich, Mario Batali, Michel Richard, Nora Pouillon, Rick Bayless, Sam Hazen, Sam Talbot, Sean Brock, and Todd Gray.

At Oceana, we are fortunate to work with dedicated ocean advocates every day. Matt Littlejohn helped us conceive the initial idea and outline. He and Emily Fisher tirelessly shepherded the book through its many stages of development. Emily also reached out to chefs and put together the wonderful recipe section. Her team, including Peter Brannen, helped with additional reporting and editing. Jim Simon, Mike Hirshfield, and Jackie Savitz asked tough but important questions about the thrust of our argument. The rest of Oceana's executive committee—Xavier Pastor, Alex Muñoz, Bettina Alonso, Audrey Matura-Shepherd, and Susan Murray—provided essential support and counsel. Margot Stiles served as reviewer/researcher extraordinaire. Cheryl Haro provided the support Andy needed to juggle writing and reviewing along with running Oceana. Jessica Wiseman and Michael Gardner got the word out as we prepared for release. In Chile, Alex Muñoz, Meredith Brown, and Paulette Dougnac helped us tell the South American portion of this story. Jon Warrenchuk opened doors in Alaska's Interior. In fact, we want to thank all of Oceana's staff, without whom our work and victories—including this book— would not be possible.

Oceana's board of directors provided guidance throughout the conception and writing of this book. Kristian Parker, a marine biologist and our chairman, and Daniel Pauly, one of the world's top fisheries scientists, cast their expert eyes over the text, and we feel more confident about our claims with their support. Oceana's other board members—Jim Sandler, Valarie Van Cleave, Simon Sidamon-Eristoff, Keith Addis, Beto Bedolfe, Ricardo Cisneros, Ted Danson, Sydney Davis, César Gaviria, María Eugenia Girón, Steve McAllister, Mike Northrup, Susan Rockefeller, Heather Stevens, Rogier van Vliet, and Sam Waterston—gave us the support and encouragement we needed. We want to thank the many friends who provided us with their guidance, including Bill Allman and, most especially, Julie McMahon. We want to thank all of Oceana's loyal funders, donors, and supporters who provide the resources we need for Oceana's campaigns.

President Bill Clinton and his team at the Clinton Foundation have been invaluable allies in the development of not just this book, by contributing a meaningful foreword, but also of Oceana's efforts to put in place effective science-based fishery management so we can save the oceans and feed the world.

We were lucky to work with a fantastic team at Rodale, whose enthusiasm about this project was something we never anticipated when we first dreamed it up. Alex Postman took our idea and turned it into reality: Her thoughtful editing and support made us better writers and collaborators. Stephen Perrine's palpable enthusiasm early on helped us realize that our book could become more than our initial modest vision. Yelena Gitlin Nesbit advised us on how to promote the book and this issue. Rebecca Klus made sure the recipes could work well for the home cook. She's got a job we think everyone must want. Finally, Kara Plikaitis created a beautiful cover and interior design, and Hope Clarke helped us get the job done smoothly.

Gena Smith and Rich Powell provided much-appreciated editorial guidance in the very early stages of the draft.

We want to thank our families. Andy thanks his wife, Beth Inabinett, who believed, with enthusiasm, it was all possible. Suzannah thanks her parents, Dan and Marilyn Evans, for their constant support, and her grandmother, Hester Shultz, for her inspiration.

Finally, while we've relied on our expert friends and colleagues for help and fact-checking, all and any mistakes remaining are our own.

ENDNOTES

THIS BOOK IS possible only because of the decades of scholarship from scientists, writers, and thinkers who foresaw the end of marine plenty long before it made the news. In just the last five years, the number of important books, articles, and reports on sustainable fishing has skyrocketed, which certainly made our job easier. This list includes many of the sources we drew on when writing our version of the story. We've divided them by chapter, but in many cases, we drew on sources in more than one chapter even if they're listed here just once.

CHAPTER 1: A SHORT NATURAL HISTORY OF SEAFOOD

Ackman, Robert G. "Fish Is More Than a Brain Food," *Microbehavior and Macroresults: Proceedings of the Tenth Biennial Conference of the International Institute of Fisheries Economics and Trade*, July 10–14, Corvallis, OR, 2001.

Crawford, Michael, and David Marsh. *The Driving Force: Food, Evolution and the Future*. New York: Harper and Row, 1989.

Cunnane, Stephen C. *Survival of the Fattest: The Key to Human Brain Evolution*. Singapore: World Scientific, 2005.

Farooqui, Akhlaq A. *Beneficial Effects of Fish Oil on Human Brain*. Dordrecht, Netherlands: Springer, 2009.

Hardy, Alister. "Was Man More Aquatic in the Past?" *New Scientist*, March 17, 1960.

McDonald, Claire. "Oldest Evidence for Deep-Sea Fishing Found." *Cosmos*, November 28, 2011.

Nettleton, Joyce A. *Omega-3 Fatty Acids and Health*. New York: Chapman and Hall, 1995.

Rabinowitch, I. M. "Clinical and Other Observations on Canadian Eskimos in the Eastern Arctic." *Canadian Medical Association Journal* 34, no. 5 (1936): 487–501.

"There's Something Fishy about Human Brain Evolution." *ScienceDaily*, February 22, 2006.

CHAPTER 2: RESERVATION FOR 9 BILLION, PLEASE

Beddington, John. *Food, Energy, Water and the Climate: A Perfect Storm of Global Events?* London: Government Office for Science, 2008.

Burkholder, JoAnn M., et al. "Impacts to a Coastal River and Estuary from Rupture of a Large Swine Waste Holding Lagoon." *Journal of Environmental Quality* 26, no. 6 (1997): 1451–66.

Centers for Disease Control and Prevention, "Heart Disease and Stroke Prevention," July 21, 2010. www.cdc.gov/chronicdisease/resources/publications/AAG/dhdsp.htm.

Community and Social Impacts of Concentrated Animal Feeds Operations. Baltimore: Pew Commission on Industrial Farm Animal Production, n.d.

Dobbs, Richard, et al. *Resource Revolution: Meeting the World's Energy, Materials, Food, and Water Needs*. McKinsey Global Institute, November 2011.

Evans, Alex. *The Feeding of the Nine Billion: Global Food Security for the 21st Century*. London: Royal Institute of International Affairs, 2009.

Food and Agriculture Organization of the United Nations. *How to Feed the World in 2050*. Rome: Food and Agriculture Organization of the United Nations, 2009.

Food and Agriculture Organization of the United Nations. *The State of Food Insecurity in the World 2010*. Rome: Food and Agriculture Organization of the United Nations, 2010.

Godfray, H. Charles J., et al. "Food Security: The Challenge of Feeding 9 Billion People." *Science* 327, no. 5967 (2010): 812–18.

Gurian-Sherman, Doug. *CAFOS Uncovered: The Untold Costs of Confined*

Animal Feeding Operations. Cambridge, MA: Union of Concerned
    Scientists, 2008.

Halden, Rolf U., and Kellogg J. Schwab. *Industrial Farm Animal Production:
    Environmental Impact of Industrial Farm Animal Production.* Baltimore:
    Pew Commission on Industrial Farm Animal Production, n.d.

Martin, Laura L., and Kelly D. Zering. "Relationships between Industrialized
    Agriculture and Environmental Consequences: The Case of Vertical
    Coordination in Broilers and Hogs." *Journal of Agricultural and Applied
    Economics* 29, no. 1 (1997): 45–56.

McMahon, Paul, et al. *What Price Resilience? Towards Sustainable and Secure
    Food Systems.* London: The Prince's Charities' International Sustainability
    Unit, July 2011.

Pimentel, David, et al. *Water Resources, Agriculture and the Environment.*
    Report No. 04-1. Ithaca, NY: New York State College of Agriculture and
    Life Sciences, Cornell University, 2004.

Schaffer, Harwood, et al. *An Economic Analysis of the Social Costs of the
    Industrialized Production of Pork in the United States.* Baltimore: Pew
    Commission on Industrial Farm Animal Production, n.d.

Sharpe, R. R., and L. A. Harper. "Methane Emissions from an Anaerobic
    Swine Lagoon." *Atmospheric Environment* 33 (1999): 3627–33.

Steinfeld, H., et al. *Livestock's Long Shadow: Environmental Issues and Options.*
    Rome: Food and Agriculture Organization of the United Nations, 2006.

Tietz, Jeff. "Boss Hog." RollingStone.com, December 14, 2006.

UK Government Office for Science. *Synthesis Report C10: Volatility in Food
    Prices.* London: UK Government Foresight Project on Global Food and
    Farming Futures, 2011.

Walker, Polly, et al. "Public Health Implications of Meat Production and
    Consumption." *Public Health Nutrition* 8, no. 4 (2005): 348–56.

Wing, Steve, and Suzanne Wolf. "Intensive Livestock Operations, Health, and
    Quality of Life among Eastern North Carolina Residents." *Environmental
    Health Perspectives* 108, no. 3 (2000): 233–38.

CHAPTER 3: SHIFTING BASELINES

Allard, Dean C. "Spencer Fullerton Baird and the Foundations of American Marine Science." *Marine Fisheries Review* 50, no. 4 (1988): 124–29.

Garstang, Walter. "The Impoverishment of the Sea: A Critical Summary of the Experimental and Statistical Evidence Bearing Upon the Alleged Depletion of the Trawling Grounds." *Journal of the Marine Biological Association of the United Kingdom* 6, no. 1 (1900): 1–69.

Jackson, Jeremy B. C. "Ecological Extinction and Evolution in the Brave New Ocean." *Proceedings of the National Academy of Sciences of the United States of America* 105: 11458–11465, 2008.

Jackson, Jeremy B. C., et al. "Historical Overfishing and the Recent Collapse of Coastal Ecosystems." *Science* 293, no. 5530 (2001): 629–37.

Jackson, Jeremy B. C. "Reefs since Columbus." *Coral Reefs* 16 (1997): S23–S32.

Kurlansky, Mark. *Cod: A Biography of the Fish That Changed the World*. New York: Penguin Books, 1998.

Marsh, Jesse. *Seafood Watch: Bluefish Report*. Monterey, CA: Monterey Bay Aquarium Foundation, October 22, 2010.

McClenachan, Loren. "Documenting Loss of Large Trophy Fish from the Florida Keys with Historical Photographs." *Conservation Biology* 23, no. 3 (2009): 636–43.

Pauly, Daniel. "Anecdotes and the Shifting Baseline Syndrome of Fisheries." *Trends in Ecology and Evolution* 10, no. 10 (1995): 430.

World Bank. *The Sunken Billions: The Economic Justification for Fisheries Reform*. Washington, DC: World Bank, 2009.

CHAPTER 4: THE CONSUMER'S DILEMMA

Abelson, Jenn, and Beth Daley. "On the Menu, but Not on Your Plate." *Boston Globe,* October 23, 2011.

Dyckman, Lawrence J. *Food Safety: Federal Oversight of Seafood Does Not Sufficiently Protect Consumers*. Washington, DC: United States General Accounting Office, 2001.

European Cetacean Bycatch Campaign. "The Tuna Boycott Which Led to the 'Dolphin Safe' Tuna Label." n.d. www.eurocbc.org/page322.html.

Food and Agriculture Organization of the United Nations. *General Situation of World Fish Stocks*. Rome: Food and Agriculture Organization of the United Nations, n.d. www.fao.org/newsroom/common/ecg/1000505/en/ stocks.pdf.

Food and Water Watch. *Import Alert: Surging Shrimp Imports Raise Consumer Health Concerns*. Washington, DC: Food and Water Watch, July 2007.

Goodnough, Abby. "A Ban on Some Seafood Has Fishermen Fuming." *New York Times,* April 21, 2012.

Lowther, Alan, Editor. *Fisheries of the United States 2010*. Silver Spring, MD: National Marine Fisheries Service, August 2011.

Molyneaux, Paul. "Protecting Swordfish and Fishers." *New York Times,* July 19, 1998.

SeaWeb and Natural Resources Defense Council. "Give Swordfish a Break!" October 3, 2002. www.seaweb.org/initiatives/swordfish/index.html.

Smithers, Rebecca. "Supermarkets Bid to Get Shoppers to Switch to Sustainable Fish." *The Guardian,* June 12, 2011.

Stiles, Margot L., et al. *Bait and Switch: How Seafood Fraud Hurts Our Oceans, Our Wallets and Our Health*. Washington, DC: Oceana, 2011.

Trenor, C. *Carting Away the Oceans VI*. Washington, DC: Greenpeace USA, April 2012.

United States Government Accountability Office. *Seafood Fraud: FDA Program Changes and Better Collaboration among Key Federal Agencies Could Improve Detection and Prevention*. Washington, DC: United States Government Accountability Office, February 2009.

United States Government Accountability Office. *Seafood Safety: FDA Needs to Improve Oversight of Imported Seafood and Better Leverage Limited Resources*. Report No. GAO-11-286. Washington, DC: United States Government Accountability Office, April 2011.

Vaughan, Adam. "Hugh Fearnley-Whittingstall's TV Fish Fight Boosts Consumption." *The Guardian,* August 8, 2011.

CHAPTER 5: CAN WE FARM OUR WAY TO ABUNDANCE?

Barrionuevo, Alexei. "Norwegians Concede a Role in Chilean Salmon Virus." *New York Times,* July 27, 2011.

Dittman, Andrew H., and Thomas P. Quinn. "Homing in Pacific Salmon: Mechanisms and Ecological Basis." *Journal of Experimental Biology* 199 (1996): 83–91.

Ewers, Justin. "In California, the Mystery of the Missing Fish." *US News and World Report,* February 14, 2008.

Ewers, Justin. "US Shuts Down West Coast Salmon Fishing and Readies Financial Aid for Fishermen." *US News and World Report,* May 2, 2008.

Fisher, Emily. "Beneath Still Waters." *Oceana* magazine, Spring 2011.

Miller, Henry M. "The Oyster in Chesapeake History." StMarysCity.org, n.d. www.stmaryscity.org/Archaeology/Oyster%20in%20Chesapeake%20 History.html.

Nash, C. E. "Achieving Policy Objectives to Increase the Value of the Seafood Industry in the United States: The Technical Feasibility and Associated Constraints." *Food Policy* 29, no. 4 (2004): 621–41.

Naylor, Rosamond L., et al. "Feeding Aquaculture in an Era of Finite Resources." *Proceedings of the National Academy of Sciences of the United States of America* 106, no. 36 (2009): 15103–10.

NOAA 10-Year Plan for Marine Aquaculture. Silver Spring, MD: National Oceanic and Atmospheric Administration, October 2007.

Northwest Fisheries Science Center, National Oceanic and Atmospheric Administration Fisheries Service. "Frequently Asked Questions: Salmon Hatchery Questions and Answers." September 14, 2006. www.nwfsc.noaa .gov/resources/search_faq.cfm?faqmaincatid=3.

Rubino, Michael, Editor. *Offshore Aquaculture in the United States: Economic Considerations, Implications and Opportunities.* NOAA Technical Memo-randum NMFS F/SPO-103. Silver Spring, MD: US Department of Commerce, July 2008.

"Sacramento Bay-Delta's Troubled History." SalmonWaterNow.com, n.d. http://salmonwaternow.org/delta-history.

"Salmon Farming in Chile." Chilean Salmon Association, August 2000.

Trauner, Carol. "Shellfish: The Positive Face of Aquaculture." *Chef's Collaborative Communiqué,* April 2004. www.ecsga.org/Pages/Sustainability/chefs_colaborative.pdf.

Thompson, Claire. "How We Can Eat Our Way Out of the Seafood Crisis." Grist.com, August 11, 2011.

Virginia Institute of Marine Science. *Virginia Shellfish Aquaculture Situation and Outlook Report.* VIMS Marine Resource Report No. 2009-5. Gloucester Point, VA: Virginia Sea Grant Marine Extension Program, June 2009.

CHAPTER 6: THE FISH WE DON'T EAT

Alder, Jacqueline, et al. "Forage Fish: From Ecosystems to Markets." *Annual Review of Environment and Resources* 33 (2008): 153–66.

Arias Schreiber, Milena, et al. "Coping Strategies to Deal with Environmental Variability and Extreme Climatic Events in the Peruvian Anchovy Fishery." *Sustainability* 3 (2011): 823–46.

Black, Jane. "Sardines Get a Modern Makeover." *Washington Post,* June 3, 2009.

"*Engraulis ringens*; Jenyns, 1842: Anchoveta." FishBase.org, July 3, 2012. www.fishbase.org/Summary/SpeciesSummary.php?ID=4&AT=peruvian+anchovy.

"Fishing in Peru: The Next Anchovy." *Economist,* May 5, 2011.

Laws, Edward A. *El Niño and the Peruvian Anchovy Fishery.* University Corporation for Atmospheric Research Global Change Instruction Program. South Orange, NJ: University Science Books, 1997. www.ucar.edu/communications/gcip/m12anchovy/anchovy.pdf.

Pikitch, E., et al. *Little Fish, Big Impact: Managing a Crucial Link in Ocean Food Webs.* Washington, DC: Lenfest Ocean Program, April 2012.

Stiles, Margot, et al. *Hungry Oceans: What Happens When the Prey Is Gone?* Washington, DC: Oceana, March 1, 2009.

CHAPTER 7: THE TERRESTRIAL TRAP

Aola Ooko, Sam. "Environmentalist? Is That a Politician with Food for Our

People?" *EcoLocalizer,* February 20, 2008. http://ecolocalizer.com/
2008/02/20/environmentalist-is-that-a-politician-with-food-for-our-people.

Butler, Rhett. "Madagascar." Mongabay.com, February 4, 2006. http://
rainforests.mongabay.com/20madagascar.htm.

"California Sea Lion: Diet and Eating Habits." SeaWorld, n.d. www.seaworld
.org/animal-info/info-books/california-sea-lion/diet.htm.

Food and Agriculture Organization of the United Nations. *Global Forest
Resources Assessment 2010.* FAO Forestry Paper No. 163. Rome: Food and
Agriculture Organization of the United Nations, 2010. www.fao.org/
docrep/013/i1757e/i1757e.pdf.

Jennings, Simon, et al. *Marine Fisheries Ecology.* Malden, MA: Wiley, 2001.

"Kill California Sea Lions?" CalCoastNews.com, March 19, 2012.
http://calcoastnews.com/2012/03/kill-california-sea-lions.

Kremen, Claire. "Traditions That Threaten." PBS.org, n.d. www.pbs.org/
edens/madagascar/paradise.htm.

Lewinsohn, Thomas M., and Paulo Inácio Prado. "How Many Species Are
There in Brazil?" *Conservation Biology* 19, no. 3 (2005): 619–24.

Manning, Rob. "Spotted Owl Surviving 20 Years after Controversial Deci-
sion." Oregon Public Broadcasting, July 17, 2012. www.opb.org/news/
article/spotted-owl-surviving-20-years-after-controversial-decision.

National Oceanic and Atmospheric Administration, Office of Sustainable
Fisheries. *Status of U.S. Fisheries: Second Quarter.* National Oceanic and
Atmospheric Administration, 2012. http://www.nmfs.noaa.gov/sfa/
statusoffisheries/2012/second/Q2_2012_FSSI_nonFSSIstocks.pdf.

Searchinger, Tim D. *The Food, Forest and Carbon Challenge.* Reston, VA:
National Wildlife Federation, December 2011.

"Upwelling and Downwelling in the Ocean." Redmap.com, n.d.
www.redmap.org.au/resources/impact-of-climate-change-on-the-marine-
environment/upwelling-and-downwelling.

Watson, Paul. "A Very Inconvenient Truth." In Defense of Animals, n.d.
www.idausa.org/essays/inconv_truth.htm.

CHAPTER 8: SWIMMING UPRIVER

Barker, James H. *Always Getting Ready: Upterrlainarluta: Yup'ik Eskimo Subsistence in Southwest Alaska.* Seattle: University of Washington Press, 1993.

Fall, James A., et al. "The Kvichak Watershed Subsistence Salmon Fishery: An Ethnographic Study." Technical Paper No. 352. Anchorage: Alaska Department of Fish and Game, Division of Subsistence, March 2010.

Joint Columbia River Management Staff. *2012 Joint Staff Report: Stock Status and Fisheries for Fall Chinook Salmon, Coho Salmon, Chum Salmon, Summer Steelhead and White Sturgeon.* Washington Department of Fish and Wildlife and Oregon Department of Fish and Wildlife, July 12, 2012.

Lichatowich, Jim. *Salmon without Rivers: A History of the Pacific Salmon Crisis.* Washington, DC: Island Press, 1999.

Matsen, Bradford. *Fishing Up North: Stories of Luck and Loss in Alaskan Waters.* Anchorage: Alaska Northwest Books, 1998.

National Marine Fisheries Service. *Table 1: Chinook Salmon Mortality in BSAI Groundfish Fisheries.* National Oceanic and Atmospheric Administration, National Marine Fisheries Service, 2012. http://alaskafisheries.noaa.gov/sustainablefisheries/inseason/chinook_salmon_mortality.pdf.

Nelson, Barry. "A Water Agenda for Governor Brown: Restoring California's Salmon Fishery and Endangered Fish." *The Switchboard,* National Resources Defense Council, November 29, 2010.

North Pacific Fisheries Management Council. *Chinook Bycatch in the GOA Pollock Fisheries: Workplan.* Item C-3(b)(1). National Oceanic and Atmospheric Administration, National Marine Fisheries Service, February 2011.

Pelc, Robin, et al. *Seafood Watch Seafood Report: Walleye Pollock.* Monterey, CA: Monterey Bay Aquarium, 2009.

Schumann, Sarah, and Seth Macinko. "Subsistence in Coastal Fisheries Policy: What's in a Word?" *Marine Policy* 31, no. 6 (2007): 706–18.

Small-Scale and Artisanal Fisheries Research Network. *About Artisanal Fisheries.* San Diego: Scripps Institution of Oceanography, n.d. http://artisanalfisheries.ucsd.edu/about-artisanal-fisheries.

CHAPTER 9: A PHILIPPINE STORY

Aburto-Oropeza, Octavio, et al. "Large Recovery of Fish Biomass in a No-Take Marine Reserve." *PLoS ONE* 6, no. 8 (2011): e23601. doi:10.1371/journal.pone.0023601.

Burke, Lauretta, and Mark Spalding. *Reefs at Risk in Southeast Asia.* Washington, DC: World Resources Institute, February 2002.

Green, Stuart J., et al. "Emerging Marine Protected Area Networks in the Coral Triangle: Lessons and Way Forward." *Conservation and Society* 9, no. 3 (2011): 173–88.

Green, Stuart J., et al. *Philippine Fisheries in Crisis: A Framework for Management.* Coastal Resource Management Project of the Department of Environment and Natural Resources. CRMP Document No. 03-CRM/2003. Cebu City, Philippines: Department of Environment and Natural Resources, 2003.

Halpern, Benjamin S. "The Impact of Marine Reserves: Do Reserves Work and Does Reserve Size Matter?" *Ecological Applications* 13, no. 1 (2003): S117–S137.

Leisher, Craig, et al. *Nature's Investment Bank: How Marine Protected Areas Contribute to Poverty Reduction.* New York: Nature Conservancy, 2007.

Lowry, G. K., et al. "Scaling Up to Networks of Marine Protected Areas in the Philippines: Biophysical, Legal, Institutional, and Social Considerations." *Coastal Management* 37 (2009): 274–90, 2009.

"Overfishing in the Philippines Threatens Whale Sharks and Local Fisheries." Panda.org, WWF International, March 3, 2006.

Pollnac, Richard, et al. "Marine Reserves as Linked Social-Economic Systems." *Proceedings of the National Academy of Sciences of the United States* 107, no. 43 (2010): 18262–65.

CHAPTER 10: HOW TO SAVE THE OCEANS AND FEED THE WORLD

We are especially indebted to the work of Prince Charles's International Sustainability Unit, which produced two excellent reports on saving the world's fisheries in 2012: *Towards Global Sustainable Fisheries: The Opportunity for Transition* and *Fisheries in Transition: 50 Interviews with the Fishing Sector,* which was the source of many of the success stories related in this chapter.

FAO Fisheries and Aquaculture Department. *The State of World Fisheries and Aquaculture 2010.* Rome: Food and Agriculture Organization of the United Nations, 2010.

Jacquet, Jennifer, et al. "Seafood Stewardship in Crisis." *Nature* 467 (2010): 28–29.

Pauly, Daniel, and Alder, Jacqueline. *Ecosystems and Human Well-Being: Current State and Trends, Volume 1.* Millennium Ecosystem Assessment Series. Washington, DC: Island Press, 2005.

Rosenberg, Andrew A., et al. "Rebuilding U.S. Fisheries: Progress and Problems." *Frontiers in Ecology and the Environment* 4, no. 6 (2006): 303–8.

Srinivasan, U. Thara, et al. "Food Security Implications of Global Marine Catch Losses Due to Overfishing." *Journal of Bioeconomics* 12 (2010): 183–200.

Srinivasan, U. Thara, et al. "Global Fisheries Losses at the Exclusive Economic Zone Level, 1950 to Present." *Marine Policy* 36, no. 2 (2011): 544–49.

SUGGESTED READING

MANY WRITERS HAVE tackled elements of ocean conservation in recent years. Here, we've listed some of our favorite reading material that offers a diversity of perspectives on the issues.

BEAUTIFUL SWIMMERS: WATERMEN, CRABS AND THE CHESAPEAKE BAY, BY WILLIAM W. WARNER (PENGUIN NATURE LIBRARY, 1994).

If you've ever smashed a blue crab with a mallet, or even if you haven't, you owe it to yourself to read this Pulitzer Prize–winning gem from naturalist William Warner. The book brings to life the peculiar personality of these crustaceans and the fishermen who catch them. It also illuminates the fragile and deteriorating state of the Chesapeake Bay, an estuary that continues to struggle against the influence of development, pollution, and a host of other land-based threats, as it appeared to Warner in the late 1970s.

BLUE FRONTIER—DISPATCHES FROM AMERICA'S OCEAN WILDERNESS, BY DAVID HELVARG (SIERRA CLUB BOOKS, 2006).

Journalist and activist David Helvarg provides a comprehensive look at America's maritime history and policies, analyzing the state of the nation's oceans by incorporating perspectives from fishermen, scientists, coastal developers, and surfers, among others. *Blue Frontier* is a definitive text in the marine conservation movement in the United States and was a catalyst for the establishment of

Helvarg's Blue Frontier Campaign and his "Seaweed Rebellion." The book's top-notch reporting and excellent writing make this a must-read for anyone concerned about saving America's seas.

BOTTOMFEEDER: HOW TO EAT ETHICALLY IN A WORLD OF VANISHING SEAFOOD, BY TARAS GRESCOE (BLOOMSBURY USA, 2008).

To put it bluntly, Grescoe is a daring eater. In *Bottomfeeder*, essentially an adventure travel book, he circles the globe to try just about every edible creature under the waves, from live octopus in Korea to barnacles in Spain. Despite Grescoe's gustatory boldness, however, the real story of *Bottomfeeder* is conservation. "The book starts off kind of adventurous and gets less and less so," he told Oceana in 2008. "But in the end, by becoming an exemplary eater, you can actually steer people away from things that are terrible for the oceans." Grescoe also reveals the inner workings of shrimp and salmon farms in scenes that are reminiscent of *Fast Food Nation*, the book that laid open the industrial processes behind the double cheeseburger. An unforgettable read.

COD: A BIOGRAPHY OF THE FISH THAT CHANGED THE WORLD, BY MARK KURLANSKY (PENGUIN BOOKS, 1998).

"Overfishing" never sounds like a very exciting topic, but Kurlansky does a terrific job bringing the long and tragic tale of North Atlantic cod to life in this page-turning bestseller. He melds the history of America with this fish in a story that will change the way you look at the early days of the country. This is the book that brought the concept of overfishing to the mainstream.

DEMON FISH: TRAVELS THROUGH THE HIDDEN WORLD OF SHARKS, BY JULIET EILPERIN (ANCHOR, 2012).

Eilperin cut her teeth writing about environmental issues for the *Washington Post*, including covering many of Oceana's campaigns over the years. Her first book is an engaging exploration of everything shark-related, from shark sex to shark conservation. She asks the big questions about why these creatures inspire everything from wonder to fear to hate in humans, without forgetting that they're under siege for their fins, the symbolic ingredient in a ceremonial Chinese soup. Sharks are awe-inspiring animals, and Eilperin's book is a fine tribute.

EYE OF THE ALBATROSS: VISIONS OF HOPE AND SURVIVAL, BY CARL SAFINA (HOLT PAPERBACKS, 2003).

MacArthur fellow Carl Safina lends a wistful touch to the story of an animal that was once nearly exterminated for its feathers but now faces an array of other human threats, from entanglement in longline fishing gear to ingesting fatal amounts of plastic debris that swirl in the Pacific. Safina's trademark lyrical style is fitting for the majesty of the northwest Hawaiian Islands he describes and the marine inhabitants that migrate to its shores.

FOR COD AND COUNTRY: SIMPLE, DELICIOUS, SUSTAINABLE COOKING, BY BARTON SEAVER (STERLING EPICURE, 2011).

Seaver was one of America's first chefs to fully embrace sustainable seafood, even if it meant he couldn't serve some of his customers' favorite fish at his restaurants. In his first cookbook, he offers up sustainable recipes along with lucid explanations of what sustainability really means, from seasonality to bycatch. He also provides a guide to a good fish kitchen in easy-to-follow instructions. This is a great place to start for anyone who's serious about serving healthy, ocean-friendly fare at home.

FOUR FISH: THE FUTURE OF THE LAST WILD FOOD, BY PAUL GREENBERG (PENGUIN PRESS, 2010).

In this James Beard Award–winning book, writer and fisherman Paul Greenberg investigates how just four fish—salmon, sea bass, cod, and tuna—came to dominate the seafood market. Through these four "archetypes of fish flesh," Greenberg explores how global overfishing has led to the collapse of wild fish populations and to the rise of aquaculture. An entertaining, vital book for anyone who eats seafood.

IN A PERFECT OCEAN: THE STATE OF FISHERIES AND ECOSYSTEMS IN THE NORTH ATLANTIC OCEAN, BY DANIEL PAULY AND JAY MACLEAN (ISLAND PRESS, 2003).

One of the world's preeminent fisheries biologists (and Oceana board member), Daniel Pauly provides an authoritative picture of life in the North Atlantic Ocean at the dawn of the 21st century. He describes an ecosystem that has been profoundly altered by decades of overfishing and offers concrete policy recommendations that would put these ailing fishing grounds on the path to recovery. For neophytes, the book provides an excellent entry point into the world of fisheries science.

OCEANA: OUR ENDANGERED OCEANS AND WHAT WE CAN DO TO SAVE THEM, BY TED DANSON WITH MICHAEL D'ORSO (RODALE, 2011).

For 2 decades, Ted Danson has worked to protect the oceans largely out of the limelight, sitting on Oceana's board of directors and acting as our voice whenever we called on him. With his first book, he gives an important and gripping overview of the challenges facing the oceans (and his own unlikely and inspiring 20+ year journey from playing Sam Malone on *Cheers* to becoming a leading, and universally respected, voice for the oceans). Danson doesn't want us to feel hopeless, however, and features a dozen "Ocean Heroes" who are doing the day-to-day work of preserving a bountiful and healthy ocean.

ON THE RUN: AN ANGLER'S JOURNEY DOWN THE STRIPER COAST, BY DAVID DIBENEDETTO (IT BOOKS, 2004).

In *On the Run*, David DiBenedetto, an editor at *Field & Stream*, illustrates why recreational fishing will continue to draw sportsmen to our shores for the rest of time. An exhilarating road trip taken by the author follows the fall migration of striped bass from Maine to the Outer Banks and the community of recreational fishermen lured to the shores to catch them. DiBenedetto also delves into the biology of the beloved fish and outlines the vicissitudes in its fortunes as it veered perilously close to extinction in the 1980s due to overfishing.

SALMON WITHOUT RIVERS: A HISTORY OF THE PACIFIC SALMON CRISIS, BY JIM LICHATOWICH (ISLAND PRESS, 2001).

In vivid prose, Lichatowich traces the history of West Coast salmon from precolonial days to the present. In the decade since he published this book, California, Washington, and Oregon salmon fisheries have descended into crisis, making his story more relevant than ever. More broadly, this book demonstrates that an ecological crisis can't be fixed with a unilateral approach. Habitat destruction, pollution, overfishing, development, and more have all played a role in the decline of the salmon. Lichatowich challenges us to a better, whole-Earth standard for protecting natural resources.

THE DEVIL'S TEETH: A TRUE STORY OF OBSESSION AND SURVIVAL AMONG AMERICA'S GREAT WHITE SHARKS, BY SUSAN CASEY (HENRY HOLT, 2005).

In *The Devil's Teeth*, journalist Susan Casey takes the reader to the otherworldly peaks of the Farallon Islands to investigate the ocean's most feared, if least understood, predators—great white sharks. It's a true adventure story of scientific research on one of the world's most inhospitable rocky outposts. Casey describes the curious history of these forgotten islands and delves into cutting-edge shark biology with equal verve.

THE END OF THE LINE: HOW OVERFISHING IS CHANGING THE WORLD AND WHAT WE EAT, BY CHARLES CLOVER (UNIVERSITY OF CALIFORNIA PRESS, 2008).

In 1990, Clover walked into the wrong press conference at The Hague. The meeting was in Dutch, but he could still understand the slides showing the devastation caused by bottom trawling on the sea floor. It took more than a decade for Clover to convince a publisher to let him tell this story, and in the meantime, he became an important journalistic voice for ocean issues at the UK's *Daily Telegraph*. *The End of the Line* became a documentary in 2009, narrated by Ted Danson.

THE MOST IMPORTANT FISH IN THE SEA: MENHADEN AND AMERICA, BY H. BRUCE FRANKLIN (SHEARWATER, 2008).

Franklin's most important fish in the sea is . . . drumroll please . . . menhaden. Most people have never heard of this bony fish, which schools off the Atlantic Coast in the billions. We asked Franklin to recommend the fecund menhaden as a potentially sustainable food fish, but he couldn't, telling us that anyone who ate the stinky fish "must have been awfully hungry." Still, menhaden have become big business. They are ground up by the ton to become everything from fertilizer to pig meal to lipstick. This industry removes from the ecosystem a critical food fish for everything from crustaceans to marine mammals to seabirds. Franklin makes a compelling case for conservation of a rarely heralded, funky little fish.

THE UNNATURAL HISTORY OF THE SEA, BY CALLUM ROBERTS (SHEARWATER, 2009).

It's impossible to overstate the contribution of this book to our understanding of humanity's impact on the marine world. Marine biologist Callum Roberts explores the belief, held by many in both the fishing industry and conservation arena, that we only started decimating the oceans since the middle of the last century with the advent of industrial fishing technology. Roberts traces our relationship with the oceans to the age of exploration and beyond. He outlines simply and effectively how generation after generation has ignored or forgotten the boom-and-bust fishing cycles of years past, and in doing so, he gives us fodder for halting the cycle once and for all.

NEW SCIENTIST

While ocean and fisheries science is often neglected by major news outlets and publications, the UK-based *New Scientist* magazine has been providing consistently thoughtful and probing coverage of the many threats and issues that impact our oceans. For those without the time to pore over scientific journals, the *New Scientist* offers a digestible, bird's-eye perspective.

INDEX